My Molar Pregnancy

*A Collection of Personal Stories
From Diagnosis Through Recovery*

My Molar Pregnancy

*A Collection of Personal Stories
From Diagnosis Through Recovery*

Edited by

Jennifer Wood

All of the medical information in this book is given either as a general overview or as individual descriptions of personal experiences. It should not be used for diagnostic or treatment purposes. For all medical concerns and questions, please contact your physician. I am not responsible for problems arising from the misuse of this material.

Most of the stories in this book were collected between September 2006 and October 2007 in response to a request posted on the Web site MyMolarPregnancy.com and relevant molar pregnancy support groups. The other stories had been previously submitted to the MyMolarPregnancy.com Web site, and their authors later granted permission for their inclusion in this book. In some cases names have been changed either at the woman's request or because there was more than one author with the same first name. Regardless, all stories published herein are owned by the women who shared them and may be used for other publications only with those women's permission. Anyone who wishes to reprint this book in whole or in part in any format should contact me by E-mail at jenn@mymolarpregnancy.com.

For additional stories, information, links, and support for molar pregnancy, visit http://www.mymolarpregnancy.com.

ISBN 978-0-6152-1225-8

For the angel in Heaven we never knew,
and the angel on Earth who came after

Contents

Introduction

I founded MyMolarPregnancy.com in June 2001, just 6 weeks after my own molar pregnancy. Confused and grieving, I had searched the Internet for information about the condition and what it meant for me in both the near and distant future. I was angry, and hurting, and jealous of all the other pregnant women around me. I needed to find someone who could understand what I was feeling and offer comfort and support. No one in my family or among my friends had ever heard of a molar pregnancy, and they could not comprehend the fear I was feeling in addition to the grief of losing a baby. After a while, many of them thought I should be "getting over it" and moving on, but I was trapped in a yearlong limbo of waiting and wondering. Would this be the month my levels go up? Would I ever get pregnant again? Was I capable of having a normal pregnancy? What was my risk of cancer, or of having another mole? At the doctor's office, the nurses who saw the elevated HCG levels in my chart assumed I was pregnant and coming in for a prenatal checkup. Time and again I had to tell them I wasn't pregnant anymore. Then I usually had to explain the entire mess all over again to convince them. My husband was supportive, but he was grieving and scared as well. He didn't want to talk about it. I felt so alone and needed so badly to share those feelings.

The first incarnation of the MyMolarPregnancy site went live on June 14, 2001. It consisted of a page or two with my story, some links and information, a guestbook, and an online support group. Within a few weeks I had nearly 20 members in the group. I soon made friends with several women in varying stages of the process. We shared our fears and grief as well as our frustration with our doctors. We learned that treatment approaches varied considerably from coast to coast and around the globe. Members began doing their own Internet research and coming to the group with new articles or information they'd found. It was the beginning of the healing process for me. I was sharing my pain with people who understood, and at the same time I was taking this horrible experience and turning it something useful that could help others. In February 2002 I purchased the MyMolarPregnancy.com domain name, and my little page became a full-fledged Web site. I began adding personal stories submitted by other women as well, because although we share a common diagnosis, and a common loss, each of us has had different experiences. Some have had chemotherapy; some have gone on to successful pregnancies. Some had their molar pregnancies long ago; some are still having their blood tests monthly; and some are just receiving their diagnoses.

I have heard so many stories, some inspiring and others devastating, but all of them familiar. This book is intended as a way of reaching out beyond the Internet to reach the women who—for whatever reason—are unable or unwilling to share their grief online but who still need the kind of support that can be found only by connecting with others who share the same experiences. Here you will find complete stories from nearly three dozen women who have been through the entire molar pregnancy experience, from diagnosis to treatment, through chemotherapy, and onward. Some, like me, have gone on to have happy, healthy pregnancies. Others have not. I won't promise you that every story has a happy ending. I can't say for certain what your future holds. But I believe that most women facing a molar pregnancy diagnosis will find a story in this book that they can relate to, and knowing you are not alone is the first step toward healing.

This book is about the subjective, personal, intimate experience of molar pregnancy. My knowledge of molar pregnancies comes mainly from my own experiences, those of the women I have met online in the past 6 years, and the research I have found online. However, it is my hope that this book will be of use not only to women with molar pregnancies but also to their doctors, their friends, and their family members. To that end, I offer the following brief description of molar pregnancies and their treatment as well as suggestions for how friends, family members, doctors, and medical personnel can best approach and support a patient with molar pregnancy to ensure the best possible outcome.

I am not a physician or scientist. The information that follows is intended to give you a general idea of what a molar pregnancy is, how it is treated, and what to expect from this diagnosis. For more in-depth, scientific, and current information I refer you to the Bibliography and Appendix of this book, where I have compiled lists of relevant articles, books, and Web sites.

Gestational Trophoblastic Neoplasia

Gestational trophoblastic neoplasia, or GTN, is an umbrella term for molar pregnancy and its related forms, including complete and partial, invasive, and persistent mole; choriocarcinoma; and placental site trophoblastic tumor.

Complete Molar Pregnancy

There are two kinds of molar pregnancy. In a *complete molar pregnancy*, there is no genetic information in the fertilized egg, so the body has only the sperm's genes from which to develop a fetus. Essentially, the sperm fertilizes an "empty egg." Nonetheless, tissue begins to grow, usually appearing on ultrasound images as black circles often described as a "cluster of grapes." I saw this "cluster" on my own ultrasound. Left untreated, the tissue will continue to grow rapidly. It triggers the pregnancy hormone, known as *human chorionic gonadotropin,* or *HCG,* and seems to feed off it, increasing HCG levels exponentially. It is not uncommon for women who are only a few weeks along with a molar pregnancy to have HCG levels in the hundred-thousands, far beyond the levels expected in a normal pregnancy. This tissue grows and grows and can, in rare cases, develop into a cancerous malignancy known as *choriocarcinoma.* Upon diagnosis, molar tissue is generally removed immediately by *dilation and curettage (D&C)* to reduce or eliminate the risk of cancer. In the case of a complete mole, there has been no actual fetus, so removal is generally done right away with little consideration.

Partial Molar Pregnancy

The second kind of molar pregnancy is a *partial molar pregnancy.* In a partial mole, two sperm fertilize the same egg. Under normal circumstances this would lead to twins, and indeed many cases of partial molar pregnancy involve twin fetuses. However, in the case of a partial mole, the fetus(es) usually has too many chromosomes and eventually dies. It's not clear why this happens. Partial molar pregnancies are especially difficult because in many cases the mother has had at least one successful ultrasound and seen evidence of a heartbeat or other fetal development. I have even heard stories from women pregnant with twins for whom one twin has been viable and the other molar, leaving these mothers with the horrible choice of whether to terminate the pregnancy or risk cancer for the sake of the viable twin. Eventually, in most cases, the partial mole is removed by D&C, although some physicians have given patients the option to wait for a natural miscarriage. In other cases, such as those involving a viable twin, a wait-and-see approach is sometimes taken, depending on the risk to the mother.

Removal and Treatment

The standard procedure to remove molar tissue is a D&C, although as I mentioned there are some cases in which a woman is given the option to wait for natural miscarriage. In other cases women miscarry naturally without knowing they have had a molar pregnancy and are diagnosed much later and only after having reproductive problems that require an ultrasound or other testing. I won't go into specifics of the D&C procedure here. However, after the procedure, which is most often done on an outpatient basis, the woman usually is permitted to go home and recuperate. Over the next few days she must have blood drawn repeatedly, and a test known as a *beta quant* will be run on the samples to measure the level of HCG in her blood. It is important that the levels decline rapidly and continue to do so until returning to normal, nonpregnant levels, generally considered to be a level less than 5. This may take several weeks, but the initial daily bloodwork will be cut back to weekly draws during that time as long as the HCG levels are declining. The woman's menstrual cycle may return during this time as well; my periods resumed 8 weeks to the day after my D&C,

but other women have had their periods return sooner and others much later.

If all goes well, the woman's HCG level should decline rapidly at first, then slower as she approaches a normal level, finally reaching normal several weeks after the D&C or miscarriage. Once a woman's HCG level reaches normal, the treatment approaches vary from doctor to doctor, region to region, and around the world. In the United States, for example, HCG is almost always measured in blood samples. However, in the United Kingdom and other parts of the world, urine samples are used. It is generally recommended that the woman be monitored for a period of time after the diagnosis to ensure that the HCG levels do not rise again. An increase in HCG would indicate that the molar tissue either has regrown or was not completely removed with the first procedure. Thus it is essential to monitor these levels for signs of regrowth. Because HCG is the "pregnancy hormone"—the hormone used to detect a pregnancy—the woman is instructed not to get pregnant during the monitoring period. This is for her safety, but it is often the source of much grief, anger, and frustration for the woman and her partner, especially if the woman's age or other medical conditions may affect her ability to get pregnant at a later time.

The standard monitoring period in the United States is 12 months for a woman with a complete molar pregnancy and 6 months for a woman with a partial mole. However, as I mentioned, this varies considerably. Some physicians make their partial molar patients wait a year, whereas others let them conceive within 3 months. A great deal of research is available online, much of it dealing with the appropriate periods of monitoring, and it is possible that with time these standards may change. However, the ultimate decision should be left up to the woman and her doctor. I do not recommend women cut short their waiting period without at least discussing it with their physicians. I have provided a list of articles dealing with waiting times in the Appendix.

In addition to blood testing, women may also be referred for a chest X-ray at the time of their diagnosis. This is not a standard procedure, but it is often done to provide a baseline image, or starting image, in the event that the molar tissue becomes cancerous and metastasizes to the lungs.

Persistent and Invasive Moles

When moles regrow after the D&C, they are considered *persistent*. Often a second and even a third D&C may be done to try to remove the tissue, and this may be successful. However, in many cases the tissue continues to grow. If the mole grows into or beyond the uterus, or *metastasizes*, it is then classified as an *invasive mole*. Persistent and invasive moles must be chemically removed with chemotherapy. Please note that being treated with chemotherapy does not necessarily mean you have cancer. Chemotherapy is designed to kill targeted tissue, and that is what is needed in this situation. Methotrexate is the most frequently used chemotherapy drug for molar tissue, but there are others as well. The chemotherapy will be administered and then the woman's HCG levels will be monitored to see if they decrease. If they do not decline, more rounds of chemotherapy with the same or another drug may be necessary. The survival rate for women with moles that regrow within the uterus is 100%, and the survival rate for those whose moles metastasize to other body parts is 97%–100% (Johnson and Schwartz 2007).

Choriocarcinoma

Choriocarcinoma is a malignant, cancerous form of GTN. It is an aggressive cancer that can rapidly spread to other parts of the body. However, it responds well to chemotherapy and has a survival rate of 75%–100% depending on the situation (Johnson and Schwartz 2007). The book *Gestational Trophoblastic Neoplasia* by Tara Johnson and Meredith Schwartz is an excellent resource for detailed information about chorio-carcinoma, the overall science behind GTN, and author Tara Johnson's personal experiences after being diagnosed with choriocarcinoma.

Placental Site Trophoblastic Tumor

Placental site trophoblastic tumor is also a cancerous form of GTN. These tumors grow inside the uterus and do not involve HCG, therefore they are detected through visual imaging procedures such as magnetic resonance imaging (MRI) or ultrasound rather than through bloodwork and HCG measurements. The tumors generally do not spread beyond the uterus, but they are not responsive to chemotherapy, thus a hysterectomy is usually performed to remove them. The survival rate varies from 20% to 100% depending on when the tumor is first diagnosed (Johnson and Schwartz 2007).

The Experience of Being Diagnosed

One of biggest complaints I have heard from women with molar pregnancies relates to their treatment by the doctors, nurses, and ultrasound and laboratory technicians involved in making the initial diagnosis and overseeing the monitoring period. In my own experience I was evaluated by a nurse practitioner who, on seeing the molar cells on my ultrasound, blurted out, "20% of all pregnancies end in miscarriage!" and then bolted from the room, leaving my husband and me in shock. She offered no words of comfort or even to prepare us for the devastating news. The ultrasound technician who did my second ultrasound was not permitted to tell us anything of what she saw, but her face was grim, and her deep, painful movements with the ultrasound wand drove the bad news home nonetheless.

Lost in a sea of confusion and misinformation, I spent 3 days alternating between hopeful optimism and crushing despair. No one even told me about the chance of developing cancer from this condition; I found that out on the Internet! As I was wheeled out of surgery after my D&C, sobbing hysterically, one nurse admonished me for my tears and told me I was young, I'd have another baby soon; in fact, I was just a baby myself (I was 26 years old at the time!). Going back to my doctor's office for checkups in the weeks that followed, I often had to explain my condition to nurses who assumed I was there for a prenatal checkup because my bloodwork showed elevated levels of HCG. None of people charged with my care seemed to comprehend the vast range of feelings I was experiencing. I had lost a baby, yes, and they had seen other patients with miscarriages before, but they looked at me as though I was being overly dramatic when I explained there was more to it, that I was facing a chance of developing cancer, and that I didn't appreciate being treated like a happy pregnant woman when I was grieving and afraid!

A woman who miscarries a pregnancy feels a devastating loss. She is left with questions and doubts: Why did it happen? Did I do something wrong? Will I be able to conceive again? Will I ever carry a baby to term? Family and friends flock around her and her partner offering sympathy and support as they grieve and begin the healing process. In most instances, the woman is told she can try again in 3 months or even sooner. Her time of waiting and healing is painful and difficult to be sure, but it passes, and the friends and family who saw her through it are still there to encourage her as she and her partner try again. It is generally expected (however unrealistically) that the partners will soon accept their loss and move onward with their lives.

For the woman who has lost a pregnancy to a mole, however, the loss of the baby is compounded by the risk of molar regrowth, the chance of developing cancer, and the requirement that the woman not get pregnant again for up to 1 year or longer. The woman in her late childbearing years who receives this diagnosis is faced with the possibility of never having children. The same may be said for women who have had multiple "natural" miscarriages and have undergone expensive fertility treatments or for those who have developed additional reproductive complications, such as endometriosis or polycystic ovary syndrome. Even for the young and otherwise healthy woman who has plenty of childbearing years ahead, the diagnosis and the required monitoring time represent a period of limbo during which she must put all of her life plans on hold while waiting to see what will happen next. For all of these women the fear of regrowth and the risk of cancer weigh heavily on their hearts and minds. Days and weeks and months of blood tests leave them feeling like pin cushions and laboratory rats. While dealing with their own fear and grief, they often also must deal with doctors who have rarely or never treated a patient with molar pregnancy, medical staff who misread their charts, and family and friends who, although supportive at the start, become confused by the ongoing "drama" of the process and gradually drift away.

I never had to face regrowth of molar tissue, because in my year of waiting, my levels never rose. It wasn't necessary for me to have chemotherapy, so I cannot begin to describe or even imagine the endless nightmare that women with persistent GTN must go through. I remember one woman, one of the early members of my support group, who had been through multiple trials of chemotherapy regimens and had lost her hair and had to walk around attached to an intravenous line. She had a young daughter as well, and her little girl had to grow up seeing her mother sick, bald, tired, and attached to strange devices. Although this woman, like many others, went on to have a healthy pregnancy, her story broke my heart and has always stayed with me. Many of the stories in this book include treatment with chemotherapy, and I will leave it to those women who know best to describe that experience.

Supporting the Molar Patient

Friends and Family

It is during this crucial period that women need support and understanding most and are least likely to receive it. Support groups like mine and others have tried to fill this void. I have come across some small "real world" support groups for women with molar pregnancies, but these are few and far between, and groups for those who go on to develop choriocarcinoma are unheard of, because the condition itself is so rare. Friends and family members often E-mail me with the same question: How can I help a loved one who has had a molar pregnancy?

Family and friends should take their cues from the woman and her partner. Some couples prefer to keep their loss to themselves, whereas others feel the need to talk. In many cases the woman wants to talk about it, but the partner does not. The best thing outsiders can do is offer support and a willing ear, but not be persistent or ask for intimate details unless the woman, partner, or couple are willing to volunteer them. Most women with GTN seek answers more than anything else, so sharing this book, and others like it, as well as Web sites that deal specifically with molar pregnancies can be helpful. Most of all, friends and family must be patient. Remember that this was not a "normal" miscarriage that the woman will just "get over." Every month for the foreseeable future she will be reminded of her loss and face renewed fear of regrowth and cancer. She won't be trying for another baby anytime soon, so don't ask things like "how's the baby making going?" or "are you trying to get pregnant again yet?" You'd be amazed how many people, even after being told about the waiting period and the risks, will nonetheless ask such insensitive and clueless questions.

Another important thing for supporters to remember is that although the woman is dealing with many things, she is primarily grieving. Her expectations of becoming a mother have been dashed. She has lost a child. And her life has been put on hold by a condition so rare it occurs in only 1 in every 1,500 pregnancies. She may desperately wish to have a baby, yet she cannot even try again for up to as long as a year. Women with GTN are left with the biggest question of all: Why me? In response, they often feel jealous of other women around them who are visibly pregnant. I can say from my own experience that when you've lost a child and are mourning that loss, it suddenly seems as though *every other woman on Earth* is pregnant except you. I remember glaring at random women on the street whose baby bumps were showing and thinking, "Why them and not me?" Coworkers and friends and family who are pregnant are even more likely targets for jealousy, because the grieving woman sees them and hears their happy stories on a regular basis. She knows it is wrong to be jealous, and in most cases she *wants* to be happy for the friend or relative who is pregnant, but the grief and jealousy can be overwhelming. Thus although the woman should not be *excluded* from baby-related events such as showers, her invitation should be offered with some sensitivity. If she declines an invitation, or does not seem to share in or listen to conversations about someone's new baby, this should be accepted gracefully. A woman who has lost a pregnancy should never be made to feel guilty or required to attend someone else's baby event, and the pregnant friend or family member should be willing to accept that the woman's absence or reticence is not intended as a personal offense.

Another difficult time for the woman with GTN is the arrival of her due date. As time passes after the diagnosis and the grief and shock wear off, things may return to normal. However, the woman has not forgotten that she was "supposed to be pregnant." Most women, on learning they are pregnant, immediately try to determine their due date. Mine was December 5, 2001; after 6 years I still remember it, and when it comes, I feel the loss all over again. The woman's friends and family can best help by remembering that date as well, at least in that first year, and doing what they can to take the woman's mind off her loss. She may well become depressed and moody, morose, or tearful. She may shut herself off from others during that time. Help her get out of the house. Don't leave her to mourn alone.

Doctors and Other Medical Personnel

I have said that women share a common complaint about the way they are treated by their medical practitioners after a molar pregnancy diagnosis. Before I go any farther, let me say this: Although I have heard many horror stories, I have also heard wonderful stories about truly caring and supportive physicians and nurses. Not everyone has a bad experience.

However, there is a common feeling among most of the women I've corresponded with that their doctors, especially their primary care physicians and gynecologists, are too inexperienced, uninformed, or regimented in their approach to molar pregnancies. The rare occurrence of GTN makes it something few doctors ever treat, and thus the "standard approach"—i.e., the one they learned in medical school years ago or found online or in a textbook after a quick search—is the only approach they can accept. Doctors generally don't like to be told they are wrong, and many are affronted by patients who bring in research or argue against "standard" treatment. Yet many of the women in my group have done just that. When faced with physicians who are unwilling to accept another approach or another line of thought, these women often take their care into their own hands, cutting short their monitoring periods to conceive again or going "doctor shopping" to find a physician more willing to negotiate. I have argued against this practice many times in my support group, urging women to follow their doctors' orders, but for the doctor–patient relationship to work, the doctor has to meet the patient halfway.

If changes to the treatment regimen are not appropriate, the doctor should be forthcoming about the reasons and the evidence that support his or her stance. There are too many emotions involved in this situation for a doctor to take a position based on the "I know best, that's why" theory. Communication and trust between the woman and her doctor are the key elements to adherence and understanding on both sides. It is time consuming (and unlikely) for doctors who rarely, if ever, treat a woman with GTN to go online or rummage through back issues of medical journals to find the latest research, particularly when there is a "standard approach" available. A patient who does her own research and brings it in to the doctor, however, has eliminated some of the legwork and offered an opportunity for further discussion, education, and experience. This kind of proactive involvement in treatment should be welcomed, not discouraged or ignored. Most of all, the woman should not be condescended to, patted on the head like a trained dog, or otherwise disrespected for her efforts. She is trying to come to terms with a situation beyond her control. Encourage, support, and be proactive along with her.

Another issue that physicians need to address when they encounter a patient with a molar pregnancy is the education of their staff and related medical personnel about the diagnosis. Charts should be clearly marked so that nurses or laboratory technicians do not mistake the woman as a "prenatal" case and offer her congratulations and a cup to pee in. A staff meeting should be held or a memo issued to staff members that describes molar pregnancy and its attendant risks. The patient should not be subjected to insensitive, uninformed, or impertinent questions. When the woman calls in for her laboratory results each month or receives a call from the office about the results, the staff member receiving or making the call should remember that the woman is waiting to hear if she might have to have chemotherapy for a regrowth or cancer. Messages and conversations should be sensitive and to the point. Friendly chatter can follow if the news is good. I had a wonderful nurse I spoke to each month at my medical center when my results came in, and having the same person on the line each time became a comfort to me, because I knew she understood what I was going through and was helping me through it one month at a time. She was my lifeline in the office, and when I encountered other staff who were clueless or insensitive, she would step in for me and clear everything up.

When I was cleared to conceive again, she cheered for me and we hugged. When I came in pregnant again months later, she was ecstatic. Confidentiality issues may make some of these recommendations impractical, but steps nonetheless should be taken to ensure these women do not feel ignored, rejected, or misunderstood when visiting their physician.

In summary, the people around a woman with GTN— whether they be friends, family, or medical personnel entrusted with her care—need to understand the wide range of emotions she is feeling and her unique needs during this time. They need to be sensitive to her loss, her grief, and her fears. They need to be patient as she struggles with her diagnosis, educates herself about her condition, and carries on through the lengthy period of monitoring and recovery. If she requires chemotherapy, supporters should take the time to learn the side effects of the treatment and offer assistance with household tasks or other responsibilities when she is too tired or sick to take care of them herself. Supporters and physicians should remember that the partner in the couple has experienced a loss as well, and is living in a state of fear for the woman's health, and they should offer whatever help or comfort the partner will accept. Physicians, particularly those with little practical experience with molar pregnancies, should be open minded and willing to go the extra mile in researching the condition when the woman seeks additional information or asks for changes in the treatment regimen. Finally, medical staff should be given the opportunity to learn more about the condition and be trained to handle any women who have had a miscarriage—regardless of cause!— with sensitivity and care.

A Word About Science Versus Emotion

Most women, when diagnosed with a molar pregnancy, will do just as I did: they will go online and look for more information. Much of what they will find will be either short blurbs such as the ones I've offered here or highly technical research articles. In all likelihood, they will find at least one article stating unequivocally that there is "no fetus" in a molar pregnancy, and therefore it is *not a real pregnancy.*

It is not my goal or even my desire to enter into a debate about what constitutes a fetus, baby, or child. However, the statement that "there is no baby" in a molar pregnancy has caused a great deal of confusion and distress to me and to many of the women I have heard from during the years. Over time, I have found that women with molar pregnancies tend to split into groups with regard to how they choose to respond to this scientific observation. At one end of the spectrum are women who choose to embrace the science. Believing that there was no baby and no pregnancy to begin with offers them a way of avoiding grief for a life lost and a means of detaching any "motherly" emotions from the situation. On the other end of the spectrum are women who have not only rejected the science but have gone so far as to name their lost child and/or bury the molar tissue, even holding services or creating memorials to the child. In the middle is the vast majority of women who simply don't know what to believe in and are left with an empty feeling that neither science nor faith can fully satisfy.

I can't offer a solution or a definitive answer as to what to believe in. I am not a particularly religious person, but I am also not a strict devotee of science. In my personal experience, when I took that pregnancy test in 2001 and it came back positive, I was pregnant. No one will ever convince me otherwise. In my heart and in my mind I bonded with the child I thought was growing inside me. I planned a future for my baby. I designed a nursery and shared my good news with my family and friends. For 4 short weeks I was a mother, and although I never saw or felt a baby, to me that child existed, however briefly. I have accepted, with the passage of time and through my work on the MyMolarPregnancy site, that there never was an actual physical life growing inside me then. But there was the spirit of a child, and the hope and dream of a child, and that is the reality I choose to live with. Whatever reality you choose to believe in, let it be the one that brings you the most comfort and the least pain.

Reference

Johnson T, Schwartz M: *Gestational Trophoblastic Neoplasia: A Guide for Women Dealing With Tumors of the Placenta, such as Choriocarcinoma, Molar Pregnancy, and Other Forms of GTN*. Victoria, BC, Canada, Trafford Publishing, 2007

Acknowledgements

This book is the culmination of a personal journey that began nearly 7 years ago with my own molar diagnosis and the creation of the MyMolarPregnancy.com Web site. In those years I have been witness to hundreds of stories, perhaps even as many as a thousand by now, that have come to me through my E-mail, my site's guestbook, and the support group. I have heard not only from women themselves, but also from their husbands, boyfriends, best friends, sisters, and mothers. I have read stories that made me burst into tears. I have been amazed at the depth of women's inner strength and the power of faith. And more than anything else, I have been proud of the way in which these women have shared their own pain and grief for the purpose of helping others endure. People are constantly thanking me for the work I have done with the Web site, but ultimately I am the one who is grateful. To all of the women who have shared their stories with me, with the group, for this book, and for the Web site, I thank you and wish you all the best of health and happiness in the future.

I also have to make a few specific acknowledgements. When I was lost and grieving, Kendra's Web site offered me a lifeline. I will be forever in her debt for that. Leslie and her daughter Gillian will always be on my mind and in my thoughts. My former bosses and coworkers were amazing to me when I was recovering from my D&C and still in the grieving and learning period after my diagnosis, especially Claire and Pam: thank you for your sensitivity, your support, and your patience. My family rushed from New Jersey to Virginia to be with me after I was diagnosed, and my mom stayed with me through the first week; I love all of you and am eternally grateful for your constant love and support. My mother-in-law, Dee, is the most amazing person I have ever met and has been through trials no person should ever experience; she is my hero and I will be forever in awe of her strength and her apparently limitless capacity for love and kindness.

I am most especially grateful to be blessed with the love and support of my husband, Jason, who endured this experience with me and kept my head above water even when he was drowning in it. You never wanted to talk about it, and I know how much it scared you, but we made it through together, as we have so many other things, and I love you more every single day. Thank you for keeping my heart and soul so very well protected.

Last, but most definitely not least, I am grateful to and grateful for my beautiful son, Xander. You are the most beautiful human being I have ever seen, inside and out. If I had not had my molar pregnancy, I would not have had you, and that would have been my greatest loss of all. I love you bigger than the sky, Sunshine.

My Molar Pregnancy

In early 2001 my husband, Jason, and I started talking about having a baby of our own. We'd both been considering it for a while, but it wasn't until that February that we decided we wanted to start trying. We knew it would probably take several months at least, so we expected we'd have lots of time to prepare, figure out our job and living situations, and start saving.

I conceived less than a month later. On March 29 I tested positive on a home pregnancy test. We drove straight to the doctor's office to have the blood test done. When the lab called back later that morning, however, the results were "inconclusive." The nurse arranged a second test, a "beta quant," to measure the amount of pregnancy hormone (human chorionic gonadotropin, or HCG) in my blood and see if I was generating more HCG than would normally be the case (a nonpregnant woman has a level of 5 or less).

We had the second test done, and the nurse called back later with the news that I was pregnant, and my HCG level was 36. We set an appointment for my first obstetric visit 3 weeks later, on April 18. This was later postponed to April 25, and the weeks went by slowly. We spent the time telling our family and friends our good news, planning the nursery, and looking into day care centers. I wanted to have everything just right when our baby arrived.

On the day of our appointment, we were excited and ready to meet our baby. Jason stood near my head and held my hand. We watched the monitor with anticipation, waiting for the first sign of our little baby. We watched, and watched, and watched. The nurse practitioner moved the ultrasound wand everywhere, much to my discomfort, but didn't say a word. Deep in the bottom of my stomach I began to feel sick. Something was wrong. "Tell me you see something!" I told her. But she couldn't; although my uterus was there clear as a bell, the only other thing she could see was a strange collection of small black circles. Without even a comforting word to prepare us, she said that "20% of all pregnancies end in miscarriage" and then practically ran out the door to schedule me for bloodwork. As I got dressed I burst into tears, sobbing hysterically on Jason's shoulder. I cried as the lab technician drew my blood. I cried all the way home. At home, we cried together.

The next day the nurse practitioner called me at work with the blood test results. The results confirmed that I had elevated HCG in my blood; in fact, my levels were up to 33,000. She told me she wasn't sure what was going on and that perhaps she'd made a mistake with the ultrasound. She wanted me to go for another at a different laboratory. I thought this meant that I could still be pregnant, that everything could be fine, that it had all been a big mistake. In retrospect, that little bit of hope was the worst thing she could have given us, because it made the truth even harder to bear.

On Friday, April 27, 2001, I went for the second ultrasound. The technicians were not allowed to tell us what they saw on the monitor; they had to tell the nurse practitioner, and she would let us know. So I lay there watching their faces as they silently did the abdominal and then the vaginal ultrasounds. I saw expressions cross their faces that I didn't want to see, didn't want to understand. They didn't have to tell me what they thought. We held out hope anyway, even as they moved and pushed and pulled the ultrasound wand until I wanted to scream from the pain and stress and emotions I was feeling. On the way home from the ultrasound, I stopped at the lab to have my blood drawn for another beta quant. The nurse practitioner called us at home a few hours later.

"The pregnancy isn't viable," she told me. She went on to tell me that they suspected I had had a molar pregnancy and that the clumps of tissue she'd seen on the ultrasound were the molar tissue. I had to have a D&C the following Monday to remove the tissue. The worst news came last. I wouldn't be able to get pregnant again for at least 12 months, and I would have to have my blood drawn regularly for monitoring. I didn't know why, didn't know what it meant to have a molar pregnancy. I turned to the one resource that had never yet failed me—the Internet. I spent the weekend searching for information. A lot of it was technical, but I am a medical editor by profession, so I was able to follow most of the jargon. Very little of it was useful to me, however, because it dealt with science, and I needed something more personal, something to help me deal with the roller coaster of emotions I was experiencing.

I finally stumbled onto a personal memorial created by a woman named Kendra. Kendra had had a molar pregnancy in 1999. She told her story in great detail, and it was there that I first learned of the potentially cancerous nature of molar pregnancies. No one to that point had even mentioned cancer, chemotherapy, or the other risks involved. All I had been told was that I'd had an abnormal pregnancy and needed to have it removed.

My D&C was scheduled for Monday, April 30, at 4:30 P.M. We arrived early and had to wait almost an hour, during which I ran the whole spectrum of emotions from grief to anger to fear. I went down to pre-op, where I changed into the hospital gown, had my IV inserted, and waited to be taken into the operating room. The doctor who was performing the procedure came and talked to me about the procedure and the molar pregnancy. I learned that my HCG level was up to 45,000 and that it was this high level of hormone, combined with the negative ultrasounds, that had caused concern and had led to the conclusion that the miscarriage was a molar pregnancy. The doctor also confirmed again that I could not try to get pregnant for at least a year and that I would have to have my blood drawn regularly for testing to ensure that my HCG levels went down to normal and stayed there. If they began to rise again, that meant the mole was regenerating. Essentially, this tissue could regrow spontaneously, and if that happened, it would have to be treated with chemotherapy.

When my time came I was put in a wheelchair and pushed down the hall. The operating room wasn't quite what I'd expected after years of watching hospital shows on television; I remember it being cold and empty except for the table in the middle and the machines surrounding the table. I got up on the table and had to move down until I was in the right position. I was cold and scared, and I began shivering and crying at the same time. A nurse dried my tears for me; I remember her face hovering above me as I drifted into sleep. I remember nothing of the procedure itself.

I came to as they were wheeling me into recovery. I was crying hysterically, knowing that it was all over. I don't know whether I'd cried throughout the procedure or not, I just remember waking up and being in hysterics. The recovery nurse was cold and unsympathetic; she kept yelling at me to stop crying and telling me that I was young, "you're just a baby yourself" she said, and would get pregnant again soon. She clearly didn't understand what I was going through, and it made me more sad and more angry so I cried harder. I only calmed down when I realized my blood pressure was sky high. I begged for Jason but was told he would not be allowed to come and see me until I stopped crying and settled down. So I lay there, sniffling, while they took my blood pressure again and again. I was bleeding, so they had to give me a pad and some freaky net-like underpants. After a while they gave me ginger ale and crackers and told me I could go home once I'd eaten and kept the food down. Jason and my mom came down to see me finally, and I was released soon after.

On the way home Jason told me that he'd spoken with the doctor after the procedure. The doctor had told him that he wasn't sure it was a molar pregnancy after all, that the tissue he'd removed didn't seem like molar tissue. He was sending samples out to be tested by two different pathology laboratories to confirm the diagnosis. In spite of everything that had happened, we felt a glimmer of hope that this was not a molar pregnancy after all, that it was just a normal miscarriage, and that maybe we'd be able to get pregnant again soon. I recuperated at home for several days but didn't hear anything from the lab. Feeling better, I went back to work on Friday, but by the end of the day I was doubled over in pain. A long weekend at the medical center and a follow-up appointment the next Monday resulted in mixed diagnoses; the after-hours clinic was certain I had an infection and had put me on antibiotics, but the doctor I saw on Monday said I was having contractions as my uterus returned to normal size. She told me to keep taking the antibiotics, however, just in case. I felt as though no one had a clue what was going on with me, and it was even more frustrating.

I had my first post-D&C beta quant blood draw on Wednesday, May 2. My HCG level by then, only 2 days after the procedure, had dropped from a high of 45,000 to 11,000. I had another draw a few days later, and it was down to 2,700. Another week and it was 1,100. Then 500, then 250, then 100. It hit nonpregnant levels (below 5) the week of June 4, 2001. On June 19 I had my second negative test, and I was cleared to start monthly testing.

The months that followed were long and hard. We had only just decided to have a baby when we'd gotten pregnant with the molar, but as we waited for clearance to try again, we became more and more convinced that having a baby was exactly what we wanted. We started planning and preparing so that we'd be more ready the next time. We also did things to take our minds off the wait; we traveled to London with some of Jason's family and spent a week in Disney World, our favorite place on Earth. We bought a new computer and did all the things, like eating out or going to movies, that we knew we'd have to put aside once we had a baby. I had a hard time seeing other women pregnant, and it seemed as though *everyone* was pregnant except me! I was jealous and angry about the whole situation. I started the MyMolarPregnancy.com Web site during this time as well, and connecting with the women in my support group helped, because they had all been in the same place I was in or were going through it at the same time. My husband refused to talk about it. He took the loss even harder than I did, I think. The combination of losing the baby and worrying about what might happen to me were a lot for him to bear, but he wouldn't acknowledge it. My friends and family were supportive at the beginning, but eventually they didn't know what to say anymore. There wasn't anything they *could* say.

Five months after being diagnosed with my molar pregnancy, I was sitting in my office in Washington, DC, editing yet another medical book, missing my baby and feeling so alone. I stopped working for a few minutes to check my E-mail, and I noticed a news headline about a plane hitting the World Trade Center. On that day, September 11, 2001, my coworkers and I were basically trapped in our office building just a few blocks from the White House while news reports came in of missing planes and the World Trade Center towers falling. The Pentagon, a few miles away, was burning. The stress and fear of that day, combined with a bad back, caused me to develop a severe case of sciatica; I could barely walk across my office because of the pain in my back and down my legs. My coworkers talked of walking into Virginia to go home rather than risk the Metro train, but I knew I could never make the 30-plus blocks it would take to get there. Sitting in my office waiting and wondering what would happen, I was actually grateful, for the first time, that I was not pregnant after all. I would have been about 6 months along by then had my baby lived. I can't imagine what it would have been like had I had an unborn baby to worry about in addition to myself and Jason. When I got home that night, and over the course of the following weeks, I came to realize that many of the things in my life that had troubled me were suddenly either forgotten or insignificant. The loss of my baby and the risk of cancer were still very much in my mind, but I realized there was nothing I could do about it. I would have to work through the grief and wait out the rest of the year, but I would use that time to get ready, and when my turn came to have a baby of my own, I would cherish every second I had with him or her. I would stop obsessing over past wrongs or mistakes and start appreciating the wonderful marriage I had and the people in my life who mattered to me.

In June 2002 I got the "all clear" from my doctor. My HCG levels had never risen once in my entire year of waiting. I had not developed cancer, had not had to have chemotherapy. I was, in short, a very fortunate woman. Yet in that year I had met so many women who did go on to develop cancer, who had lost their hair or had multiple D&Cs. I also met women who went on after their molar pregnancies to have happy, healthy babies. I hoped for the same for myself.

On a Saturday morning in late October 2002, as Jason slept, I took yet another pregnancy test. It was our fourth go-round with trying to conceive. As the second line appeared on the test, clear as a bell, I burst into tears. I was pregnant again. I climbed into the bed next to Jason, and when he woke he thought at first that I was crying because I *wasn't* pregnant yet. He soon learned the truth, and we held each other and started dreaming, once again, about the baby to come. This time, however, our joy was coupled with an avalanche of fear. Would it happen again? Could we handle losing another baby? We swore right away that we would keep this pregnancy to ourselves until the end of the first trimester and until we had confirmed the baby was healthy.

The next few weeks were very long and difficult. Instead of cherishing and enjoying my pregnancy, I spent every day in terror. A blood test indicated that I was pregnant—there were no "inconclusive" tests this time!—but I was anxiously awaiting my first ultrasound, which was scheduled for mid-November, at roughly 7 weeks. When that day came, I felt a terrible confusion of emotions: fear mixed with excitement, anticipation mixed with dread. These feelings worsened when the test came back with unclear results. They simply couldn't see anything yet for certain. The doctor tried to reassure me by saying that it was probably just too early, but I was horrified nonetheless.

Our second ultrasound was scheduled for the day after Thanksgiving. With the weight of fear and the secret of our pregnancy pressing on us, it was a less than happy holiday. We muddled through the family gatherings, putting on happy faces and trying desperately to keep everything inside. By the time we got to our appointment the next day I was ready to burst. It seemed so unfair that a period in my life that should have been joyful and wondrous had been overshadowed by so much fear and uncertainty. That was about to change for the better, however. On that day, November 29, 2002, we saw our baby's heartbeat for the first time, flickering away on the screen. At long last, we had a baby.

I'd love to say the rest of my pregnancy was doubt-free and perfect, but although I tried to cherish every second, the fear persisted. A side effect of running a support group like mine is that you hear every horror story possible. I knew of women who'd lost their babies to moles as late as 22 weeks. So I never really got over that initial terror of something going wrong. Nevertheless, I did what I could to savor the experience of being pregnant. We began making plans for our future family. At Christmas we shared our news with our parents and began to enjoy the celebration of a life to come. In February 2003 I had another ultrasound, and we learned not only that our baby was healthy and active but also that it was a boy! We were having a boy! We began thinking about names. In March we decided to leave the DC area where we were living and move to Florida, where we could buy a house and possibly live on one income. I wanted to be a stay-home mom and have a house and a yard for our son to play in.

We found a house in April and made plans to move in sometime in the summer. I was due on June 30, so we figured we'd stay in DC until the baby was born, and for a few weeks after, and then make the move. My husband is a teacher, so we had to schedule the move between the day he finished at his old school in Virginia and the day he had to report to his new school in Florida. We started packing things up before I would be too big to help, but we took our time, because we figured we had weeks, even as long as 2 months, yet to go.

My baby shower was held in New Jersey on March 18, 2003. Driving home to Virginia that night, I felt completely exhausted. I was so wiped out I couldn't go to work the next day. Over the next few days, things began to happen that worried me and made me believe I was in labor, but the doctors told me I was just having some stretching and some Braxton-Hicks contractions. They told me I had weeks to go, not to worry. On Thursday, May 22, I told Jason to get ready. I told him it didn't matter what the doctors said, I could tell things were happening. The baby was coming, and soon.

Our son was born the next morning, May 23, 2003, at 7:08 A.M. My water had broken in the middle of the night, and at 6 A.M. the doctor determined the baby was in breech position and had to be delivered by emergency cesarean section. Despite his early arrival, our son weighed a whopping 6 pounds, 1 ounce, making him the heavyweight champ of the hospital's neonatal intensive care unit, where he stayed for a week before coming home. We finally had the family we'd wanted for so long.

After my son was born, a family member told me she'd never seen any mother who *appreciated* her child as much as I did. I know what she meant. I cherish every second of my son's life, even the difficult and frustrating moments (and there have been more than a few!), because we went through a nightmare of loss and fear before we could get to where we are today. But the point is, we made it.

Alwynn

The day my husband and I got married, all I could think about was wanting to start a family. My career wasn't where I wanted it to be, but I didn't care; like so many other things in life—I wanted it NOW. My husband wasn't so sure, however; he wanted to be more financially secure. There were problems with his company, and he was worried about losing his job. So we decided to wait. I was rather surprised, then, to realize—after he had was laid off from his job—that I was pregnant. Worrying about money put a damper on things, but we were happy nonetheless. Then, a few weeks later, he had a new job, and everything seemed to be perfect!

I was so excited to tell my mother about the baby; she had always wanted to be a grandmother, and although she had a beautiful stepgrandson already, this was something she had been hoping for from my sister and I since we were born! We knew it was premature, but we began hunting for baby things and stockpiling them away. I started reading all sorts of books, and my husband and I laughed as we thought up baby names.

Suddenly, on a beautiful day in August—just after I had told a few people at work that I was pregnant—I went to the washroom and was horrified to see thick, brownish blood. I really think I felt my heart stop. I immediately sent a quick note to my boss, jumped in my car, and went home. I called my doctor, who said if the blood wasn't bright red, I should not worry about it; all I could do was wait and see. My husband did his best to get home from work as soon as he could, because I was nearly hysterical. I kept telling myself that this happens to some women, everything was okay, I wasn't going to lose this baby. Days went by and the blood didn't stop. I called the hospital several times but didn't get any help, so finally I told them I was coming in for an appointment. My husband did what he could to change his schedule and take a few days to be with me, hoping his new job would understand.

I spent a long, miserable day at the hospital. My doctor was on vacation, so I saw one of her colleagues, who was the doctor of a close friend of mine, and she was wonderful. They couldn't find the source of the bleeding, so I was sent for an ultrasound. We had to wait a couple of hours before the appointment. All the while I felt like I was going to throw up from the anxiety. The woman who did the ultrasound was very pleasant and informed me that the radiologist needed to come in to consult. I couldn't hear much of what they were saying other than "placenta." Never "baby" or "fetus." I knew something was terribly wrong. When I went back up to see the doctor, she told me that there was a fetal pole, but it didn't seem to be more than about 6 weeks or so. They were also worried it could be a molar pregnancy, which would mean I would have to have a D&C. She told me I should go home, and she would call me later to discuss what we would do next.

The doctor called me that night and told me I would have the D&C the next day. She had consulted with a colleague and thought it best to err on the side of caution by having the D&C rather than inducing the miscarriage. My parents came down early the next day so they could take us to the hospital. Although my day at the hospital was fairly quick, it was far from painless. The staff at the hospital were wonderful, but each and every moment that passed I felt swept over by the most mind-numbing despair. I could only hope that they were wrong, that it wasn't this "molar" pregnancy (I hadn't done much research on this yet; I was hoping I didn't have one because I had heard I would have to wait 6 months to try to conceive again!) and that with time and some physical and emotional healing, we could try again soon.

I recovered well from the surgery and convinced myself that surely I couldn't be one of these rare women who has a molar pregnancy. Then I got a call at work confirming the diagnosis: Yes, it was a partial molar pregnancy, and I'd need to come in for a follow-up appointment. I would have to continue doing blood tests every week. I couldn't believe it. Hadn't we been through enough? Now I had to worry about a cancer-like condition and could not try to get pregnant again for months to come? I almost completely shut down. I left my office again, sobbing almost uncontrollably, thinking that if my luck continued, surely I'd end up with this type of cancer, and things would never get better. My mother came down again to be with me and help me get through this because my husband had to return to work.

I went diligently for my blood tests every week. Just 4 weeks after my D&C I had my first negative result. Thank goodness! Maybe there was a bit of light at the tunnel after all! I researched as much as I could about this condition and found the MyMolarPregnancy site as well as a few others. I found message boards where I could read other women's experiences and share my own. I couldn't believe how much this helped. Talking to women that I didn't have to explain everything to made my life so much easier. It was on one of these message boards that I read about a paper written by researchers from the New England Trophoblastic Disease

Center who claimed that with a partial molar pregnancy, women whose levels fall to negative in 7 weeks or less only had to be followed up for 3 months. Eureka! This was the first good news I'd had in ages. I printed off the paper and left it with my doctor, who admitted she had not dealt with molar pregnancies much in the past. In fact, she had had only one previous molar patient, a woman with a complete mole who had had no desire to have any more children. My doctor consulted with the specialist who had performed my D&C. They both considered the study authors to be the "gurus" of trophoblastic disease, so they agreed that I could try to conceive again in 3 months.

I was so thrilled. My "sentence" had been cut in half! I would be able to start trying again in December. Thus it came as a shock to me when, at the end of *November* I found out I was pregnant again. My husband and I realized that we had had ONE slip up in all the time since we were told to wait, but apparently that one time was enough. We were terrified at first, but I had to believe that fate couldn't be so cruel as to do this to me twice. However, my husband had been let go from his new job during that time (apparently they weren't as understanding about the time he'd taken off as we had thought!). Regardless, we were determined to be optimistic.

My doctor was very understanding and told me to come in each week to have my levels checked to make sure they were okay. We would do an early ultrasound at around 6.5 weeks to see how things looked. I was so excited. Then, while I was Christmas shopping with a friend, I started to get horrible cramps. I told myself that women sometimes get cramps when they are pregnant, but then suddenly I saw blood. No, no, no! It was NOT happening again! Everything seemed to be falling apart around me. I spent another day in the hospital, and all the memories and emotions from the molar pregnancy came flooding back. They couldn't find the source of the bleeding, so I was sent for an ultrasound. As I lay there during the ultrasound, I wondered if I would ever be there for *positive* reasons. When she was done the radiologist said, "I'm afraid your pregnancy is ectopic. It's about 2 cm in diameter and is in your left fallopian tube. You'll need to take a note back up to your doctor, and she can discuss what you should do."

Come on! You're kidding me, right? A molar pregnancy AND an ectopic? How is this even possible? But it was. I was sent off to see a gynecologist, who decided the best course of action was to give me an injection of methotrexate. As a "folic antagonist" it destroys the folic acid in your body that the embryo needs in order to grow. When the embryo stops growing, your body reabsorbs it, and that's that. So now I had a 1% chance of having another molar pregnancy again (not so bad) and about a 12% chance of having another ectopic (not so great). I'd have to come in again each week to follow my levels down to zero (gee, why did that sound familiar?).

Six months after my D&C—and a little more than 2 months after the ectopic pregnancy—I was cleared to try and conceive again. I was terrified, but I was not going to let that stop me. I admit I'm not a particularly strong person of faith. But I believed I had had enough pain and was due for some better luck. If nothing else, I knew that I certainly could *get* pregnant!

On October 30, 2004, one day after my 28th birthday, I gave birth to a beautiful 6 lb, 9 oz baby boy. I had complications during my pregnancy; my alpha-fetoprotein test came back abnormally high, and the doctors were concerned about possible spinal issues. It turned out to be a placental defect that, in my case, didn't cause any serious problems for the baby. It did, however, cause my blood pressure to skyrocket just shy of 2 weeks before my due date, so I had to be induced early. Unfortunately, my body didn't progress well, so after 48 hours of labor, I ended up having a cesarean section. I was in the hospital for almost a week, because on the day we were to go home, they realized the baby had very low blood sugar. He had to be admitted to the neonatal intensive care unit and put on an IV as well as be treated for jaundice. Thankfully, 36 hours later we were able to take him home.

Holding him in my arms finally, after all the loss, pain, and uncertainty, was the most incredible moment of my life. Every day is the most amazing gift and I realize that I would go through anything, even all of this, 100 times over to have him. If I were to offer advice to anyone who has had a molar pregnancy, it would be this: You are your own best advocate. I have heard so many stories of women not being followed properly, getting conflicting advice, and so on. Do your homework, write down questions for your doctor, and if you don't think he or she is doing everything possible to help you, find a new doctor. Your health, and possibly the health of your future children, is too important.

Caren

I found out I was pregnant in April 2003 at the age of 31. When I told my boyfriend that I was pregnant, I wasn't sure how he would react. His response could not have been better; he told me, "I couldn't be happier. This is the most important thing that has ever happened to me." Although unexpected, the pregnancy was welcomed by our family and friends as well, which made what happened next even more painful.

We went to hear the heartbeat when I was 11 weeks pregnant, on May 21, 2003. I had been overwhelmed with anxiety that something was wrong, and I was abnormally fearful of miscarriage. I thought that hearing the heartbeat would allay my fears. However, the doctor was unable to locate a heartbeat. I was immediately sent for an ultrasound. The technicians said that the fetus looked to be about 7 weeks, with no detectable heartbeat, and that the placenta was abnormally shaped. That is all I remember.

My doctor called me later that day and said I had had a missed abortion. I was scheduled for a D&C 2 days later. The procedure was not painful or physically taxing, but it depleted me emotionally. In the days after the D&C I was filled with a feeling of doom. I felt like I would never have children. I felt uncertain of the existence of God. How could a loving God allow this to happen to two people who were so eager to be parents? I felt like I was being punished because I had had an abortion when I was 24. To add insult to injury, my doctor called 2 weeks later and told me I had had what was called a *partial molar pregnancy,* and it would be necessary to come in for blood tests on a weekly basis.

My HCG levels dropped very quickly. They were 37, then 9, then negative. I knew I should be thankful that I had not developed other possible complications, but I was angry and resentful that I had had to experience this enormous disappointment in the first place. I was depressed and discouraged about my ability to have children. Oddly, the wife of a coworker had the same thing happen to her only a few weeks before it happened to me. I wondered, was this caused by environmental conditions? Or was it just an unbelievable coincidence?

During my waiting time my boyfriend and I became engaged and were married in October 2003. I was cleared to conceive again a month later, and by December 10, 2003, I was pregnant again. Our baby is due August 24, 2004. So far, I have seen the heartbeat three times and all is progressing just as it should. I have been scared and anxious during this pregnancy, but as time passes I am filled with more joy and anticipation. I have bought maternity clothes. We have chosen names. I have hope again.

Caroline

I was 25, and my fiancé was 30, when our first pregnancy ended in miscarriage around the 11th week, in September 2001. It came as a shock, because we had seen a healthy-looking fetus with a strong heartbeat on our first ultrasound at 7 weeks. What followed was the saddest, most painful and upsetting period of my life. I spent days—weeks—crying tears for my lost baby. The world seemed like such a horrible place to live in. How little did I know at that time that there was still more to come.

A week after the D&C—which also was my first surgery ever—my job committed me to move abroad, away from my fiancé. After we got married in December 2001, I decided to quit my job and move back to live with my husband in The Netherlands. It was all so wonderful and exciting, and in January 2002 I felt that my body as well as my soul had had enough time to recover from the trauma of the miscarriage. I was ready to take a new, deep breath of fresh air.

We tried again to conceive, and right away my period came late. I ran to the drugstore for a home pregnancy test. It was positive! My husband and I were very excited but tried not to get out hopes up too high. I set an appointment with my gynecologist for February 28, 2002. Throughout February I felt very nauseated, and when my breasts stopped swelling around February 20 I sensed that something was wrong. But

I tried to keep believing that I could still be pregnant because I was still experiencing morning sickness.

On February 25 I discovered a brownish discharge that freaked me out. I called the hospital, but they advised me to remain calm and wait for the scheduled appointment. Those were the longest 3 days of my life. I knew something had gone wrong and was upset that I couldn't bring myself to feel a connection to the baby inside of me anymore, even though it might still be healthy and growing. It made me feel like a traitor. I thought if my suspicions about the baby were to be wrong, and I was feeling this way anyway, I didn't deserve to be a mother.

On the day of the appointment, which was supposed to be my 8th week of pregnancy, I was waiting anxiously in the doctor's office, hoping with my last strength that everything would be all right, that all the suspicions and anxieties were just due to my hormones. Finally, when it was my turn, the gynecologist took us to the ultrasound room, and indeed there was no fetus, no heartbeat, just some unusual placenta. My husband started to get teary, and I felt like my heart was burning out and falling into ashes, but that was just the beginning. Back in the waiting room, a feeling of emptiness overcame me as I sat there, with all the pregnant women and new moms, filling out an admission form for yet another D&C. Why does such a cruel event have to happen to a woman at all? I felt that if there were a God, he surely didn't love me, and if this was a punishment, I at least wanted know why. That day I began to lose all faith I ever had inside of me.

Four weeks after the D&C I was called back to the doctor's office, where I was told that the lab results indicated a partial molar pregnancy. This was apparently very rare in Holland. The gynecologist could not explain to me what exactly it was, and I got the feeling that she was trying to avoid me. She just told me that I would have to monitor my HCG level (which at that time was at 144,000) weekly; avoid getting pregnant for the next 6 months; and start taking the birth control pill when my period returned. I was scheduled for another appointment a month later, in May 2002. I left the hospital confused and feeling like a horrifying loser, without even knowing *why* exactly. I decided it would probably be best not to talk about it with other people.

At the appointment in May, I told my gynecologist that I was concerned because I still had not gotten my period, even though it had been nearly 3 months since the D&C. She told me it was a side effect from the molar pregnancy and not to be worried, because my HCG level had gone down to 16,000. I left the doctor's office still confused about my condition. Fortunately, my period finally arrived in June, and although the colors were strangely darker than usual, it seemed to be normal, so I started my birth control pill.

My next two periods were a total nightmare. Blood clots—lots of them—kept coming out, for 2 weeks at a time! I felt weak and even fainted on a couple of occasions. I called my gynecologist, but she told me to remain calm and wait, because it could just be a side effect of the molar pregnancy. By then it had been 5 months since my D&C! My husband and my mom became concerned about my condition and decided to take me to see my mom's gynecologist, which meant I had to travel halfway around the globe to Bangkok, Thailand! There the gynecologist was very concerned that my HCG level was still at 146 almost 5 months after the D&C. He did another ultrasound, which showed that molar tissue had grown back in my uterus. He had my lungs x-rayed to make sure that they had not been affected and scheduled me for another D&C the following day. A week later my HCG level was down to 2, but my overall blood test indicated anemia, and I had to start taking medication for that.

The gynecologist was puzzled why my gynecologist in Holland didn't take any action earlier. He explained molar pregnancy to me and also advised me to look it up on the Internet—how I wish I had thought of that earlier! But then, I had never really been sure what to look for, because my previous gynecologist had always avoided any conversation about molar pregnancy. Now that I finally was aware of my condition, I found the strength to get the word out to our families, but their reactions were not what I expected. Besides my husband, my parents, and my grandmother, everyone else seemed to want to keep their distance (my in-laws even tried to blame me and make me feel bad about having a molar pregnancy and "bad genes"). Nevertheless, I tried to understand everyone's reactions. My sister-in-law had given birth to her second son in May, and my aunt was expecting her third baby in October, so perhaps people preferred to keep their distance from "bad news" that might be "contagious."

Yet this experience was quite painful. I wouldn't have made it through without my husband and my dog. I felt repulsive and in shame. I thought it better not to talk about it to anyone anymore, until a couple of months later, when the topic happened to come up during a conversation with close friends. They were very concerned and seemed interested in what exactly a molar pregnancy was. That positive event gave me confidence to talk about it to a couple of other close friends, who were also amazingly helpful. This finally gave me the strength to cope with the whole experience. I began to read just about every single article and story on the Internet and joined two lifesaving online molar pregnancy groups, which gave me faith and hope for a possible happy ending after all.

My blood was monitored monthly for almost a year before I got the green light to try to conceive again. We had recently relocated from The Netherlands to South America, back to my husband's home country, when we learned we were pregnant again. My husband and I were cautious about announcing the pregnancy; it was hard for us even to believe that we could have a normal pregnancy. However, this pregnancy was uncomplicated; I experienced no morning sickness at all, just some minor headaches during the fourth and fifth months and a little bit of swelling toward the last month. It was a true joy, and I couldn't wait to meet the baby girl inside me. It was also bittersweet, because I sometimes wondered how big the other two babies would have been by then.

In February 2004, our daughter was born at a healthy 49 cm and weighing 3,300 grams. She has grown up to be a very lively little girl and the true sunshine of our lives. I pray and thank God every day for this blessing.

Dana

Before I found out I was pregnant in June 2003, I noticed how bloated I looked in my brother's wedding photos. I felt a little nauseated and really didn't want to have many drinks—which, in retrospect, is odd, because I tended to enjoy having two or three socially, especially at weddings.

We had just purchased our dream home. It was summer, and our house had a pool, so we decided to take the summer off to have friends over and enjoy lazing around and having cocktails on the deck. A week after we moved in, we had some friends over for the afternoon. I had a few drinks but was feeling very tired and somewhat unsettled. I was a week "late," and my breasts were sore in a way unlike my usual menstrual tenderness. My friends then convinced me to take a pregnancy test, which I thought was ridiculous. It was instantly positive. I had never had a positive test, so the fact that it became positive so quickly and very brightly so soon in the pregnancy didn't send up any red flags at the time. After the first test, we decided to go to the store and get more tests to confirm. As we walked out my front door, I got sick and vomited in our front yard! I was so embarrassed and hoped the neighbors didn't see that display.

My husband and I were very surprised and a little taken aback. We had discussed having children when we moved into our home, but had expected to enjoy the summer and get into serious talk about kids after.

Later that week we went to my doctor to confirm the results and discuss what to do next. She performed a vaginal ultrasound and she saw TWO sacs with what seemed to be two fetal poles. We were beside ourselves. I had always had a feeling that I would have twins, and it was coming true! At that point, my HCG was around normal for a multiple pregnancy.

We were stunned but happily the accepted news. I can honestly say that the next weeks went by without incident. I didn't notice any exceptional nausea or spotting; nothing that could have alerted me to what was going on in my body. I walked around our lake every morning before work and began the healthy lifestyle of a pregnant woman.

At the next appointment, my doctor frowned when she performed the ultrasound. She was unable to find a heartbeat, and something didn't look "right" with the sacs. We were 9 weeks along, and the heartbeat should have been obvious. I began crying immediately. I had never wanted children until I met my husband, yet here I was, devastated by the possibility that I could be having a miscarriage. The doctor was comforting and mentioned half-heartedly that she could be wrong. She then sent us to the hospital next door for a detailed ultrasound just to be sure.

After hours of pacing in the waiting room for the ultrasound, we were brought into the cold examination room. As the technician performed the exam, she was unconsciously making odd faces. I asked what she saw, and she seemed uncomfortable and stated that she was not allowed to comment on the results, that we had to wait for the radiologist. After the technician finished, we were left waiting in the room for what seemed to be an eternity. The radiologist finally came in and confirmed that we had miscarried. There was no elaboration, just that the pregnancy wasn't viable. I was beside myself. We were told to go back to my doctor to discuss our options. We could either have a natural miscarriage, which would drag out the inevitable for some time, or have a D&C. I chose the latter so that I could get the ordeal over with and move forward.

The D&C was routine and uneventful. I was sent home with instructions not to try to conceive for at least one menstrual cycle and to return for a follow-up appointment in a few weeks.

The follow-up was the point at which our life got a bit scary. I had never heard of a molar pregnancy before. My doctor told me that I had had a partial molar pregnancy and that I would have to come in for weekly blood draws until my HCG levels went down. I honestly thought nothing of the diagnosis at that point. My doctor didn't seem to be particularly concerned or alarmed, so I assumed it was a somewhat normal condition. I called my mom on the way home from my appointment and told her the situation. She searched on the Internet after our conversation to learn more about what was going on. About an hour later she called me back very upset and sent me several articles indicating that this wasn't common and could be quite serious.

Luckily, my levels fell to zero within 1 month, and we were given the green light to try to conceive after my next cycle. I breathed a huge sigh of relief that day, realizing that all of the frightening things I had read about weren't going to happen with me. That this was just a blip in my life and that chapter was closed for good.

About 3 weeks after my last blood draw, I felt kind of strange (I cannot explain why, other than just intuition). My husband and I had not had sex since the D&C, so I couldn't quite figure out why I decided to buy a pregnancy test. It was instantly, blazingly pink. I started crying and knew something was wrong. I called my doctor and went in the next day for a blood draw.

My HCG levels were over 150,000. I had experienced what my doctor felt was virtually impossible, a recurrent partial molar pregnancy. There were SEVEN moles present in my uterus, and a D&C had to be performed immediately.

My life became a whirlwind at this point. I was starting to grieve the loss of my babies now that I had bypassed all of the consequences of the molar pregnancy, but that had to be put on hold again—I had to concentrate on my own life for the moment. After several weeks of blood draws, my HCG blood count was still not regressing. The anxiety I felt every week waiting for the nurse to call with the results was enough to send me through the roof! When she called the day that my levels increased, I was devastated. I was at my office and could not control my tears. Why me? Why after so much would my body not just let go of this pregnancy?

My doctor then sent me for chest x-rays and a battery of tests to determine if the molar cells had spread to any other part of my body. I was an emotional wreck. The thought that these possibly cancerous cells were coursing through my blood and could infest other organs was overwhelming. I didn't sleep until the results came through. My husband, luckily, could keep a level head and was able to calm me during the wait. Fortunately, no metastases were found—all of the cells had been contained in my uterus.

In order to bring my HCG levels back to zero, I was put on a short regimen of methotrexate to knock out any other cells that were present in my system. Chemotherapy—wow. I knew it was a very light and "easy" form of therapy, but it was still very traumatic. I bought the drug in the pharmacy downstairs and brought it to my doctor for injection. I was relieved that I did not have any adverse reaction from the treatment other than exhaustion.

Weekly blood draws were sent out by my doctor as "stat" because my husband and I couldn't stand the normal 2-day wait for results. After a few treatments, my HCG levels fell to zero. The weight that fell from my shoulders that day was immeasurable. We went to dinner with friends, and I enjoyed my first margarita in months—and I was loopy. I will never forget the toast: "This too, has passed" stated one of my friends. Although she was right about the medical aspect of this disease, I was still reeling emotionally from the roller coaster we'd been on.

One of the hardest parts of all of this for me, however, was that I had not had time to mourn the loss of my babies. I had had to deal with *my* health first. When the medical scare was over, I went into a deep depression. I would go out drinking with my friends only to come home and have horrific breakdowns in front of my husband. I couldn't figure out why; I was healthy and had been told there was virtually a 100% chance that I would go on to have a normal, healthy reproductive life. It then occurred to me that I hadn't really said goodbye, and I hadn't given myself the chance to be sad that I miscarried. So my heart was doing it 6 months later, once it could handle the emotional weight of it all. On the babies' due date I held a personal vigil. I wrote them a letter and planted flowers.

My husband and I began therapy together to figure out how I could move forward, even though it seemed my problems were now in the past. It truly helped discussing my feelings and pain with someone who was objective. My husband was a blessing, but he was unable to stand back enough emotionally to be my crutch. Within a few months I had learned to deal with the heartbreak and loss without losing grip on my current reality. I could finally say that I had put that past year behind me and move on with my life.

We were given the go-ahead by my doctor to begin trying to conceive again in April 2004, and we were able to get pregnant that same month. The mixed emotions of becoming pregnant were staggering. I was petrified for the ENTIRE pregnancy that something was going to go wrong, that a molar was going to occur again, and that I would lose yet another chance to be a mother. However, after a mostly uneventful pregnancy, our angel Aislin was born in January 2005. The post-pregnancy routine seemed pretty similar to any other birth, with the exception that my placenta was sent to a lab to determine whether there were any molar cells present. I went in for my 6-week postnatal exam and was relieved to find that everything was normal—no molar cells to be found.

It is hard for me to believe that this whole molar saga could be over. I guess I will never be completely free of the worry that my life could change so quickly like it did. For now, I am enjoying my second post-molar baby girl, Carys. I feel blessed to know that these are my miracle babies and that I have the chance to love them each moment of every day.

Eileen

I had been going through infertility treatments for 18 months and was finally proclaimed pregnant with not one but two babies. Wow! We were so excited.

At 6 weeks we saw both sacs by ultrasound. At 7 weeks, we were told that one of the embryos was not viable but that the second had a healthy heartbeat. We were sad about the loss but understood that these things sometimes happen. At around 8 weeks, I had a severe migraine, but I went to my obstetric appointment, and the doctor said everything was fine. One week later we went in for a third ultrasound. We found out that we had lost the second fetus sometime around the eighth week, the same timeframe as that migraine. We were devastated.

I was scheduled for a D&C 3 days later but went in for a final ultrasound the day before the surgery. We wanted to be sure that this was a definite miscarriage. The D&C was performed on March 8, 2002. I would have been almost 10 weeks pregnant. My doctor, being proactive, had my HCG beta checked 1 week later. It was above 99,000. He was horrified and rushed me into his office and did another ultrasound to see if any tissue had been left behind or if, because it had been a GIFT (gamete intra-fallopian transfer, an infertility procedure), there was possibly a third, ectopic pregnancy. The technician mentioned something about a mole, unbeknownst to me at the time, but according to the hospital there was nothing found in the tissue. My doctor, still very concerned, sent the labs off to his primary hospital, where the pathologists detected a partial mole.

I was told to have weekly blood tests to monitor the beta HCG, and I was sent to an oncologist (a member of the International Society for Trophoblastic Disease) who said the same—we just needed to wait. If the levels went down, we followed it, if not, we took action.

Four months later (just about 17 weeks), my beta was still at 592; higher than it was with my first pregnancy test. Everyone kept telling me to wait. I couldn't, much longer. My research had indicated that the HCG level should be back to less than 5 within 8 weeks *at the most*. Each week I went in, I relived the loss of the child that my husband and I had wanted so badly. Each week that my level was up meant a longer wait for us to try again and another week that we did not know what would happen in my body.

One week later, I went in for my weekly bloodwork and got the devastating news that my numbers had gone up to well over 700. My doctor and I both called my oncologist, but he had been in England for the previous 3 weeks and would not be back until the following Monday. I did not know what to do, who to call, how to react. In essence, I had just found out that I had cancer. The oncologist on call said not to worry, that there was no emergency. She said chemotherapy was likely, but we would have to wait until Monday. What did she mean, don't worry?! Not an emergency?!

Well, I did not have to wait until Monday. That Sunday morning, I woke up hemorrhaging. I called the doctor on call, again, and she said if I went through a pad an hour to call back. I went through a pad in 40 minutes and was experiencing incredible pain like nothing I had ever known. I called back, and she said to go to the emergency room. We were an hour and a half from my hospital, so my husband stopped halfway there and picked up some Advil; I couldn't stand the pain! I took four, because I had been told this was like a muscle relaxant.

We got to the emergency room and they took me within 10 minutes, although it felt like hours. My blood pressure had skyrocketed, and I was still bleeding. They brought me to a bed and took my blood pressure again. It had gone down somewhat, probably from the Advil, and my pain had been eased at least a little. I felt something and asked to go to the bathroom. There I passed something that looked like small balls or grapes. Scared, and I guess out of habit, I flushed the toilet. The doctors were disappointed but thought that I may have passed the tumor. They proceeded to do a lung scan and a chest x-ray and told me that I would be admitted. My doctor would be back the next day, and we would start chemotherapy. The next day they woke me at 5 A.M. just to say "hi." They ran tests all day and finally began a 12-hour course of methotrexate at 8 P.M. At 8 the next morning I was taken off the IV and given antinausea medication, antibiotics, and instructions. I was released around 1:00 P.M.

One week later, I had my first post-chemo bloodwork. My beta HCG was less than 5. Bloodwork was done every week for 5 consecutive weeks and then monthly for 1 year. I was finally cleared in August 2003 and allowed to continue trying to conceive. We had one unsuccessful embryo transfer and one unsuccessful GIFT procedure. We then went to a second clinic, where they reevaluated our situation. We went through two in vitro procedures and one timed intercourse. Five years and 12 procedures later, we have decided to stop trying the medical route to parenthood. When we started fertility treatments, everything I read said that when it was time we would know. It is time. We are now exploring the adoption route. Sometimes God has other plans for us. We were dealt lousy cards, but we have to play the hand through.

Fran

My name is Fran. In 1985, after having four children, I had a hydatidiform pregnancy. I was 21 weeks when I went in for my first ultrasound. I hadn't had much trouble with my other pregnancies, and I wasn't expecting anything wrong. I took my two preschoolers with me. They were in the room with me when the technician blurted out, "Sorry, this baby is dead." I then somehow managed to drive us home. A few days later I went to the hospital. I was told I needed to go through labor to expel the baby. I hemorrhaged so badly that one of the nurses was completely covered with blood. The doctors had everyone come into the room to see and learn about the molar pregnancy. Most of them had never seen one before. I felt like I was the show-and-tell item of the week. Because I had lost so much blood, I had to have a blood transfusion. I also had a D&C. After a 2-day stay in the hospital, I went home and was told it was okay to try again.

I got pregnant the same time the next year. At 20 weeks I had the dreaded ultrasound. The doctors called this one a "missed miscarriage," where the baby dies but stays in the uterus. Again, I went through labor and then had a D&C and another blood transfusion.

In 1987 I again was pregnant. I switched to a high-risk obstetrician. This time, I was really sick. It was summer, so I was hot, and I couldn't eat anything but a tiny bit of baby food. My 10-year-old took care of my other children for me, because I couldn't do much of anything. They lived that summer on pizzas and packaged cookies, because that was about all she could fix. Every time I ate or moved I would be sick. When I was 6 weeks along I needed maternity clothes. When I was about 10 or 12 weeks along the doctor said it looked like twins and that was why I was so sick. At 14 weeks I couldn't fit into my maternity clothes and had to get bigger sizes. We had promised the kids a trip to Disneyland in August for being so helpful. I didn't feel well, but I wanted to see my family in California before the twins came. I started to dream about double strollers and everything else that goes with twins. I was so excited, because I had wanted twins since I was a little girl. At Disneyland I stayed in a wheelchair the whole day and got a really good perspective on how we sometimes treat people in a wheelchair.

We came home from our trip, and the next morning I woke up bleeding. I was at 22 weeks. I went in to see the doctor, and he confirmed another hydatidiform mole. The problem was that there was still one baby with a heartbeat. He scheduled surgery in about 5 days so we could get everything ready. I am very religious and against abortion, and here I was, faced with a must-have abortion. It was a soul-wrenching time. I prayed that the baby would die in the next few days so I wouldn't have to kill him. The doctors were afraid that I would hemorrhage and die during the surgery. I spent those days playing with my children and getting their Christmas gifts ready in case I wasn't around. When I went into the hospital the baby was still alive. I was started in labor, but complications arose and I had to have an emergency D&C. This was so rushed that I had no pain medication. Then I delivered the baby and had another D&C. They took chest x-rays and monitored my blood for a year.

I was told that if I got pregnant again there was a 50% chance of a molar and only a 20% chance of a healthy baby. I decided to call it quits and had my tubes tied. I was also told that because I had had so many moles, my children might also have them, if they could even get pregnant. I worried about this for many years, but I am now the proud grandmother of three 2-year-olds.

At the time this happened I wanted a support group, but I was told that I hadn't really lost a baby, they were just miscarriages. I would have loved to bury the last baby, but I was told then that it wasn't possible. I'm glad this information is now available.

Heather

On New Year's Eve 2005–2006, we were bringing in the new year with great excitement. We had just announced to family and friends that we were expecting. Thoughts of upgrading to a bigger place, registering for nursery items, and having our upcoming ultrasound were dancing in our heads.

After returning home from a New Year's bash, I was ready to settle in for the night, tired as ever. To my concern, I had a bit of spotting. Immediately I got on the phone and called my obstetrician. At this time, I was 9 weeks pregnant. He suggested that I go to the emergency room for some observation.

When I arrived, I was immediately taken back. I had some bloodwork sent to the lab and an internal exam done to check for complications. After a few hours of sitting in the emergency room, a nurse told me that I was being sent for an internal ultrasound. The results from the ultrasound showed some irregularities. When the results of the bloodwork came in, the doctor on call entered my room. I could tell by the expression on her face that something was wrong. I had already begun to prepare for the possibility of a miscarriage. I had no idea it would be more complicated than that. "It's a molar pregnancy," she explained. Molar pregnancy? What's that? Millions of questions were running through my head.

I was admitted to the hospital and sent to the labor and delivery unit. My obstetrician came in and answered a lot of our questions. I had been diagnosed with a complete mole, and immediate surgery was necessary. I went into surgery on New Year's Day. My surgery went well. There was no major blood loss or damage to my reproductive organs. My doctor then explained that I would be sent for weekly and then biweekly blood tests to monitor my HCG levels. As long as my HCG levels were dropping, this meant that my body was ridding itself of the abnormal tissues.

On returning home, my focus was on getting healthy and getting through the next few weeks. Others were focused on the fact that I lost a baby, but I suppose I was still so scared about all the "what if's?" that I hadn't mourned the baby's loss.

My HCG level started at 250,000 before surgery. The week after surgery it had dropped to 5,000. My numbers slowly dropped. Then, the call came that my HCG levels had plateaued. Not a good sign. I was going to be referred to an oncologist. Oncology...I knew what that meant. I was frantically searching the Internet for explanations and other women's stories. My husband was extremely supportive. He never let me know how scared he was. Only later did he share his fears and sorrow surrounding these things. He was so positive and reassuring that I could really lean on him through these scary times.

I went the next day to the oncologist's office. I met with my doctor, who told me that due to a persistent mole, I now had what was called "gestational trophoblastic neoplasia." The surgery had evacuated as much tissue as possible, but the remaining tissue was not flushing itself out. It was persistent. Chemotherapy would be necessary. I sat in the doctor's office with my husband, crying and fearing the worst.

So many scary thoughts run through your head. Will I lose my hair? Will this medicine make me really sick? Will I have to stop working? What if the medicine isn't effective? I was beyond terrified. The worst part was that nobody I knew had ever been through this or anything like it. I felt so alone.

I scheduled my first appointment for that Friday. At that appointment I met with a nurse who showed me a notebook that explained in detail all of the possible side effects, how the chemotherapy drug would be administered, and much more information. I was also asked to choose which drug I wanted to use to treat the disease. Wow. I couldn't believe that I had to choose.

It was narrowed down to two drugs: methotrexate and actinomycin-D. I was given information to look over about each drug, including side effects, methods of injection, and other important things. It seemed as though there were fewer serious possible side effects with methotrexate, and I found more studies and information about it. After consulting with my doctor, we decided that would be the treatment drug. I followed the chemo nurse back into the treatment room and received my first injection. It was an intramuscular injection into the buttocks. My nurse warned me that some feel a burning sensation or have pain at the injection site, but my experience was not bad at all.

I proceeded to have weekly injections for 10 weeks, which is considered to be five rounds of chemo. Each round consists of two injections and lab work after the second injection to monitor HCG levels. My levels dropped quickly at first, but some of the tumors took longer to get rid of.

During the treatment, I felt pretty well. I was very tired, as could be expected. I did not lose any hair or have any thinning. I was not nauseated and did not have any vomiting or diarrhea. I did have very dry and burning eyes. I also had rectal spasms for a week or so that were very painful. I ate a lot of fiber to soften my stools, which ultimately stopped the spasms.

My last round of chemo was in April 2006. I have been getting bloodwork done each month to monitor my HCG levels, and my numbers have remained negative. Yeah! Each time I get bloodwork, it is a little scary. But it's also more encouraging each time I go and that dreaded phone call doesn't come.

I started to feel the loss more once the whirlwind slowed down. Each day is different. I am confident that I will have children. I have been told by my doctors that there should be no problem getting pregnant and having a happy, healthy baby. I have to wait 1 year from my last chemo injection to start trying again. I'm already counting down the days.

To anyone who might be going through these same experiences or something similar: You will get through it. Stay on top of your treatment and your bloodwork. Do not wait. Be proactive and stay in constant communication with your doctors. Ask a lot of questions. The more you know, the less scary it seems.

Thank you for reading my story. I hope it has helped.

Jane

My husband and I have two boys ages 11 and 6. For 2 years we had been trying to decide whether we should have another child. I have high blood pressure, so we were not sure it would be a good idea. My doctor told me to be satisfied with two children. Finally I asked my obstetrician, and he said my last two pregnancies went okay, and we would deal with the blood pressure problem. Also, my three best friends had recently had their third babies, so I was feeling a desire to have a baby too. We tried for 6 months, which seemed long; it had only taken 1 or 2 months with our first two children. Of course, I was much older this time, at 36, than I was before. We found out through a home test that we were pregnant on December 29, 2003. We were both still unsure at this point but excited. We decided to keep it a secret until after the first appointment at 12 weeks. We didn't even tell our boys. I was a little leery because with my second pregnancy I bled for a weekend, and we thought we miscarried. By some miracle we didn't.

From the start of this third pregnancy it seemed different. I had a lot of cramping, was very tired, and the morning sickness started immediately and sometimes lasted all day. I attributed all of this to being older, 20 pounds overweight, and out of shape! I went to my first doctor's appointment on February 6, 2004, which was just to see the nurse to fill out the paperwork. I was not scheduled to see the doctor until February 18. About a week before this I had had some spotting, which seemed to be mucus tainted with blood. At the appointment, my blood pressure was elevated, so the nurse spoke with my doctor; he was between appointments at the moment, so he chose to do an ultrasound right then. As he did the scan, I knew something was wrong because he wasn't talking to me. He said things didn't look right, and he wanted to do a vaginal ultrasound. He left the room so I could get undressed. I said a prayer that I would see the heartbeat. He did the ultrasound, then told me he thought I had had a molar pregnancy and began to explain the bad news. I had never heard of this. It was all very surreal.

I didn't cry at that point. I think I was in shock. I just remember wanting to leave as soon as possible. I somehow found my car and burst into tears. I was supposed to only be gone from work for about an hour. I teach school, so not being there on time is a problem. I drove home, and luckily my husband was home. I explained what I could and then had to return to school because there were no substitute teachers available. I pretended everything was okay for the rest of the afternoon. I went on the Internet that day to learn more about molar pregnancy and learned way more than I wanted to know! This was a Friday. The nurse called me on Monday to confirm that my HCG levels were too high: 125,000. A D&C was scheduled for Thursday. On Tuesday I had to see the doctor for another ultrasound to confirm one more time that it was the right diagnosis. After school on Wednesday we learned that one of our teachers had died that morning unexpectedly. Thursday morning I had the D&C and, in the doctor's words, "It went fantastic." That definitely was not how I felt. Friday I went to the wake for my friend and on Saturday to his funeral. I think I was still numb from all of this. I don't know if I can quite separate the emotions of the two events. I kept telling myself that things could be worse; my friend no longer had her husband! But that didn't make the pain of the molar pregnancy go away.

My HCG levels dropped to 23,000 right away, then 13,000, then 3,300. I had a chest x-ray done as well. My blood pressure didn't come down, so I had to go see another doctor to change my medication. I couldn't use regular birth control pills because of the blood pressure problems, so my obstetrician had me to try an alternative. All of these tests and doctor appointments were overwhelming. Depression set in, and it felt like I was alone in this. I was worried about cancer and upset about not being able to try to get pregnant again for a year. I felt like I was running out of time. I regretted not having a third child sooner. I also found it difficult to explain to friends, relatives, and coworkers why I had to have a blood test each week. I don't think the receptionist at the clinic even understood.

I continued to have blood draws until October 2004. My HCG levels dropped to zero, and I got through it all without developing cancer. My obstetrician then said I could try to get pregnant again. We conceived again in January, and I told my husband the good news on his 40th birthday, February 9, 2005. I had my first doctor appointment on March 1, 2005. There was a heartbeat; however, the doctor said the fetus didn't seem to be developed as much as it should be for the number of weeks I was pregnant. I had to go back in 10 days. When I went back on March 10, there was no heartbeat, and I was miscarrying. After a discussion with the doctor, it was decided that a D&C would be best. We couldn't believe this was happening again. It was determined that this was just a miscarriage and not a molar pregnancy this time. That was a relief, because I didn't want to do all those blood tests again and worry about cancer. My doctor told me that 1 out of 3 pregnancies will result in miscarriage in women my age. I will be 38 in June. We can try to get pregnant again now, but I am so unsure. I keep praying for a sign from God!

Janna

I was diagnosed with a partial molar pregnancy on March 18, 2003.

My husband and I had waited 5 years before starting a family. I got pregnant in November 2002. This was our first pregnancy, and my husband and I were eagerly anticipating the arrival of our baby in August. I was in my second year of teaching special education, we owned a home, we had two dogs…. It was all "picture perfect" and going according to plan. We told our family and close friends in December that we were expecting. It was a wonderful Christmas present, and our parents were so excited to become grandparents! I even learned that my college roommate was also pregnant, and we were due on the same day!

I began experiencing extreme nausea at 7 weeks. Because this was my first pregnancy, I didn't know what to expect. I did my best to manage, but each week the nausea and the fatigue just got worse. I had an obstetric appointment at 11 weeks. I wanted one thing and one thing only at that visit: to hear my baby's heartbeat. My doctor told me not to hold my breath, because the Doppler didn't always pick up the heartbeat that early. I went back a week later, and still no heartbeat on the Doppler. My doctor told me not to worry and to come back in 3 weeks. I wasn't worried, because I was really nauseated and fatigued—all good things, I thought. At 13 weeks my nausea in the morning had all but disappeared. According to all the books I read, I was doing great!

By 15 weeks I had gained weight and appeared to be "showing" a little. I went in for my doctor's appointment, positive that I would finally get to hear the baby's heart. My husband and I met at the doctor's office. We were both excited. Needless to say, the excitement turned to fear, and my doctor sent me across town for an ultrasound. I had such a sinking feeling when I walked into the radiology clinic. My husband, ever the optimist, held my hand and told me not to worry. I knew that there was something wrong, and I sat in the waiting room feeling jittery and anxious. After waiting for what seemed like a lifetime, the technician called me back and got me situated on the exam table. The technician started the ultrasound, and right away I knew there was something wrong. Even to my untrained eye, I could tell that there was no heartbeat. At 5 P.M. on February 28, 2003, the ultrasound confirmed a fear that had crept into my mind, but that I didn't want to believe. Our baby had passed away.

A still, lifeless form showed up on the monitor. The technician said the baby measured 8 weeks. The technician called my doctor on the phone. We sat in the ultrasound room, crying, while we waited to speak to the doctor. A lot of what happened afterward that is a blur. I remember thinking how horrible it was going to be to tell everyone that we had miscarried. I felt sad, embarrassed, and overwhelmed. I was referred to an obstetrician for a D&C (My doctor was a Family Practice/Obstetrics doctor). My new doctor was a gift from heaven. She was full of empathy, and she put my mind at ease. I had surgery on March 4, 2002.

After surgery, it felt like my heart hurt worse than my body. My husband and I grieved for the baby that we were to have held in our arms at the end of the summer. However, we were surrounded by an outpouring of love and prayer from our friends and family, which helped our emotional healing. I was off work for a week after the surgery. When I returned to work, everyone was so empathetic. Ironically, one of my close coworkers asked me, "It wasn't a molar pregnancy, was it?" I told her no, and I remember thinking, "what the heck is that?!?"

Two weeks after surgery, I went back to the doctor for my surgical follow-up. That's when the rug was pulled out from under me. The nurse escorted me back to the exam room, and after the routine questions, she said "The doctor will be right in to explain your pathology." I still didn't think anything was the matter until my doctor walked in. I took one look at her face, and I knew something was wrong.

She pulled her chair up right next to me and told me that the pathology report indicated a partial molar pregnancy. My coworker's question rushed to my mind, but I didn't know what the implications would be to me! My mind went into slow motion as I listened to the words come out of her mouth. Words like *metastasize, cancer,* and *chemotherapy* were not supposed to be used in reference to me! To her credit, my doctor treated me with amazing understanding. I asked her to repeat a lot of what she said, because I couldn't believe what I was hearing. How could all of this come from a miscarriage? She wrote me a prescription for birth control pills, explained to me that I couldn't become pregnant for a year, and sent me off to the hospital for a chest x-ray and a beta quant count. I remember walking out to my car in a complete daze, and by the time I reached the hospital I was in a panic. My husband was at work, and I couldn't reach him on his cell phone, so I sat alone in the waiting room of the hospital radiology department just crying and crying.

Thus began a 2-month journey of blood tests and waiting for results. I don't know what hurt the most, being scared for my health or having to wait for at least another year to start our family. My chest x-ray was clear, so I never had to repeat it. My first HCG count was 176, which was relatively low for a molar pregnancy. My levels dropped slowly each week, but they were dropping! On April 23, 2003, my levels hit zero. It was truly an answer to a prayer, because I had "stalled-out" at 8. One more week and my doctor was going to give me the methotrexate shot.

Once I was given a clean bill of health, I started to take stock of my situation. Waiting an entire year to get pregnant again seemed like an eternity. Thankfully, I believe that God brought other things into my life to occupy my mind. I started taking better care of myself and lost nearly 30 pounds. I took a different teaching job closer to home, which gave me a focus and a new challenge. My husband and I also drew closer to each other in a situation that could have torn us apart. That year was one of the most challenging, yet rewarding, years of our life together.

Now, more than 3 years later, I am blessed to say that a sweet little boy is sleeping just down the hall as I write this story. Little did I know that my molar pregnancy experience was just the beginning of a long, hard journey. I would love to say that the molar pregnancy was just a hiccup, and everything went fine after it. Unfortunately, I went through several more painful miscarriages after the molar pregnancy and the long 1-year wait. Finally, in December 2005, I became pregnant. An early ultrasound confirmed that I was pregnant with twins, but as I neared my 12th week of pregnancy I miscarried one of the babies. Although I was sad, we were still very thankful for the one healthy, developing little baby. I was due on August 25, which ironically was the same due date as my molar pregnancy! I relished my pregnancy and enjoyed it to the fullest.

On August 1, I began experiencing symptoms of what I thought was stomach flu. By the next morning I felt pretty ill, so my husband called my obstetrician (the same doctor who had guided us through the molar pregnancy loss), and she asked me to go into the birth center, "just to make sure I wasn't dehydrated." Unfortunately, by the time I reached the hospital that morning I had developed a rare and life-threatening form of severe preeclampsia called HELLP syndrome. In less than 2 hours my liver was in peril, my platelets had dropped dangerously low, and I had developed severe hypertension. My doctor recognized the symptoms immediately, and I was rushed into an emergency cesarean section. Our son Carter was delivered at 1:55 P.M. on August 2. He weighed 5 lbs 12 oz and was 1 month premature, but he was healthy and spent only one night in the neonatal intensive care unit. After several harrowing days in which I was given magnesium sulfate and one platelet transfusion, I began to recover from the effects of the HELLP syndrome, and we left the hospital as a family. There is no known medical reason as to why I developed HELLP, and my doctor said there is no correlation between the molar pregnancy and HELLP syndrome.

Obviously, our road to parenthood has not been an easy one. For most women, fertility and pregnancy are things that come naturally and easily. I don't know why I experienced the molar pregnancy and all the subsequent medical issues, but I know that it has made me a stronger person. Through the challenges, God has been so faithful to my husband and me. We know that everything happens for a reason. I received excellent care from the obstetrician that I was referred to "by chance" when I had my molar pregnancy. That very same doctor literally saved my life and the life of our son 3 years later. The molar pregnancy has caused us to be grateful in a way that we wouldn't be had we not gone through it. Now, as I reflect on the entire journey, I am thankful for the experience, because it has made the victory all the sweeter. There is truly life after molar pregnancy!

Jeana

In August 2003 my husband and I got married. He was 41, and I was 25. He had been married before, but they were not able to have children, so we decided that we would try to get pregnant right away. I had always wanted children. We tried for 3 months without result, and my husband kept saying after every negative pregnancy test, "Don't worry, honey. We'll keep trying."

In October 2003 I was finally late on my period, and 2 days later I took a pregnancy test. It was very faint, but there were two pink lines. I rushed over to show my mom. My husband and I were so excited, because we had finally done it. We were getting our baby! Being first-time parents, we called everyone we knew to tell them that we were having a baby. We started buying clothes and baby furniture, and I made a bumper and blanket for the crib. The thought of miscarriage never entered our minds. I went to my first appointment at 7 weeks, and everything checked out fine. The nurse midwife had us set an appointment to hear the heartbeat at 11 weeks.

It had never occurred to me that we wouldn't be able to hear our baby's heartbeat. To me, the hard part was *getting* pregnant, not staying pregnant. We went in with the video camera, all excited. The nurse midwife tried for 20 minutes to find the heartbeat, but we could only find mine. She told us not to worry and had us make the appointment for the next week. We went home, and I fell apart. I told my husband that I knew the baby was dead. He told me that everything would be okay. I also called my parents and told them the same thing. For the next week, I was on pins and needles.

Next week finally came, and the birthing center called to ask if I would change my appointment to an earlier time. I did, and my husband was not able to go with me, so my mother did. Once again we brought the video camera, and once again the nurse midwife could not find the heartbeat. She sent me for an ultrasound about 30 minutes later. We went over, and the technician had me get ready for a vaginal ultrasound. There was my baby, perfect, with fingers and toes. I looked at him and said, "He's not moving." Sure enough, there was no heartbeat. The baby measured at 9½ weeks when he should have been 12 weeks.

I was devastated, and I called my husband. He left work immediately, as did my dad. All I could do was cry. We talked to the nurse midwife, and she told us that it would be best to wait for me to go into labor naturally. I went home in a daze and walked into a house that was waiting for a baby. I burst into tears and had my mom and dad take out all of the baby stuff. They took it to their house and stored it until I could use it again.

All of this took place on December 17. For the next week I felt like I was sitting on a time bomb. My husband drove me every day to my parents' house, and then he would drive to work. He usually drove in a company truck with another guy, but he drove his car so that he could leave as soon as something started. By Monday, I was a nervous wreck, so we made an appointment to see the doctor. He saw me that day and scheduled a D&C for the next morning, December 23. My parents and my husband waited in the waiting room until it was all over. The doctor came in and said everything went well and that they were going to send the tissues to the pathologist. He wanted me to see him again in 2 weeks.

The holidays were ruined for our whole family, but we tried to enjoy it as much as possible. I had a follow-up on January 6. The doctor had previously told us that they probably wouldn't be able to tell us what caused the baby's heart to stop beating and that if everything looked good, we could try again after a couple of normal cycles. I went in, and the doctor did his exam. He said that everything looked good and asked me how I was feeling. I told him I was looking forward to trying again. That was when I was informed that I had had a partial molar pregnancy and that I had to go in for blood tests once a month until my HCG was back to zero, and then for 3 more months after that.

Once again, the baby we wanted was beyond our grasp, and I was a 25-year-old facing the possibility of cancer. My husband was very positive and supportive. I couldn't have gotten through it without him. Reading other women's stories helped me, because the doctor couldn't tell me much about molar pregnancies.

It took about 2 months for my HCG levels to go down. After that, it seemed like every month the doctor would tell us that we had to wait another month to start trying again. We wanted a child so badly that we even started the process of becoming foster parents. In April 2004 I went to another doctor to get a second opinion and was cleared to start trying again. We believe in divine healing, so I also got anointed and was prayed for at our church.

I got pregnant again in June 2004, and now have a beautiful daughter named Aubrianna Austyn. She was born on March 6, 2005, weighed 6 lb. 6 oz, and was 19¼ inches long. My husband is in love with her. The pregnancy, labor, and delivery were uneventful.

The Bible verse that I would like to share is found in I Samuel 1:27 and reads, "For this child I prayed; and the Lord hath given me my petition which I asked of Him." The advice I'd like to share is *don't give up*. Don't give up on hopes, dreams, family, and friends. Most important, don't give up on God! I have been there. This is one of the most devastating things that you can ever hear, that your child is dead and that you may never be able to have another one. But somehow you live through it, and life goes on. I still can't look at the baby book I started for our first child without crying. It may always be that way, I don't know. I hope this has encouraged you as other stories did for me. Good luck with everything and God bless!

Jennifer

I was 32 years old when my husband and I celebrated our 4-year anniversary in September 2005. We had been together for 10 years, traveled often, and led a very fortunate life. We were not sure when we were in our 20s whether children would be a part of our life. We both had blossoming careers and loved our life together. As we entered our 30s, however, things just sort of changed for us. We felt we were missing something. We talked about children for a few years, and around that 4-year anniversary in 2005 we decided to start trying to conceive. Many people we knew had had some problems conceiving, so we thought we would give it a few months. Well as luck would have it, that first month was all we needed. In October 2005, the home test quickly showed two lines indicating that I was pregnant. In an instant, life had changed. At my first prenatal visit at 6 weeks, the doctor conducted the basic tests and we saw the first ultrasound. There it was, a pretty non-descript sac, but clear as day we all saw the heartbeat. The doctor said that everything looked okay and told us come back in 4 weeks.

We had planned a family Thanksgiving vacation and decided that by then I'd be about 10 weeks and it would be a good idea to tell the whole family. We did and told them that we were still keeping it quiet until the next visit, which would be after we returned from vacation. I felt really good; I had some soreness in my breasts but very little morning sickness, and I was still jogging, although I was trying to take it easier. I thought that I probably should be having some food aversions or nausea, but some people never have any pregnancy symptoms, so I thought nothing of it. The family was really excited; I tried to contain my mom's excitement, telling her that I really wanted to wait until the 12-week mark before getting too excited myself. I knew of so many friends who had had problems, including ectopic pregnancy and miscarriage, that I wanted to wait until the nuchal test before being overjoyed. My mom, who luckily never had a pregnancy-related issue herself, continued to encourage me and had her heart in the right place when she repeatedly told me not to worry. I, however, am a pessimist by nature; I tend to expect the worst so that I'm pleasantly surprised when things work out and yet never disappointed. I took the same approach here. Expect the worst, I thought, because things always surprise you and work out okay.

The Wednesday after we returned from vacation I had my next visit. The doctor conducted a standard ultrasound, this time from my belly instead of transvaginally like the first time. When she couldn't find the heartbeat she didn't think much of it, telling me that it was still early, and then she tried scanning transvaginally. That didn't ease my anxiety or my husband's fear on his face. We just sort of knew, without saying a word to each other. The rest was a blur, because as soon as she looked at the monitor after the transvaginal, she had that look—she turned that monitor toward me and said, "I'm sorry, there is no heartbeat." Those words reverberated in my mind for months afterward. I cried, asked what I had done—was it the glass of wine I had with dinner that one night on vacation? Was it my running? Of course, none of those things "cause" a miscarriage, but finding a reason, especially for me, was paramount.

After we gathered ourselves, the doctor explained that it looked like the fetus had died about 2 weeks earlier, based on the size. Because nothing was happening naturally, she recommended a D&C for 2 days later. I agreed; I couldn't stand the idea of carrying around something that would never be. I have had major issues with medical procedures in the past—basically, I faint when blood is taken or during a routine gynecological exam—so imagine my fear of a D&C! I threw up twice that night when I returned home, making myself sick with terror and sadness, and I couldn't believe this was happening. The conversation with my parents was heart wrenching. My mom was flabbergasted that this really happened to her daughter. She was consumed with worry for me and my ability to deal with this, as I myself and my husband were! This was going to be a terrible road ahead.

After the D&C on December 9, 2005, which in retrospect went really smoothly, I had a 2-week follow up visit. At that visit the doctor recommended that, with a miscarriage, we wait about 2 months before trying to conceive again. She also mentioned that a pathology was conducted on the products of conception from the D&C—a standard procedure for her and the hospital—which gave me some comfort because I was still searching for a reason. She warned, of course, that most pathology reports don't give conclusive reasons for the miscarriage and that most often it is deemed as just bad luck.

On January 19, 2006, 4 weeks after my follow-up, I had another check-up. It was then the doctor explained that the pathology laboratory reported I had had a partial molar pregnancy. A what? It sounded like a dental condition. She explained that two sperms fertilized one egg and that the fetus had 70 chromosomes (23+23+23+1), where one of the chromosomes replicated again. She said it happened at the moment of conception and that there was absolutely nothing that we did to cause it, nor did it indicate any genetic or hereditary concern—it was just bad luck on the order of 1 in

1,000 to 1 in 1,500. Pretty uncommon, but not totally rare. I actually felt better momentarily—at least there was a clear reason. But then she told us that we would have to wait 6 months before trying to conceive again.

What? Six months? If they knew what it was, why such a long wait? Then came a flurry of worries. Monthly blood tests, cancer, HCG levels, chemotherapy…. We just wanted to have a baby. How could all this be possible? I researched as much as I could on the Internet. It was all so confusing. I was devastated, and my "expect the worst" attitude was not helping in this situation. I was convinced that I would develop choriocarcinoma—cancer—and that I would have to go on chemotherapy. Although I knew the cure rate was close to 100%, I'd still be 35 by the time we could try to conceive again. Funny how in my 20s being a mom had been just a fleeting thought, and now it was the only thing I could imagine myself being. My amazing husband did all he could to try keep me focused on one day at a time—no easy task.

I came to realize that the women who were having the most difficulty (e.g., rising HCG, infertility issues, chemo treatments) were also those seeking the most support from others online, and thus it was difficult to find a story from someone whose experience was positive. When things go well, people seemed less likely to seek support and to share their stories. Being unable to find positive stories I found it difficult to believe that I could have a positive outcome. But we took each day as it came—we had no choice. I needed to feel like I was taking some control, so with my great "type-A" personality I set up a spreadsheet schedule of blood tests, projected ovulation dates, and potential dates when we could try to conceive again.

My first blood test on January 19, 2006, showed an HCG level of 13. My doctor was pleased and said that even after a miscarriage *not* related to a mole, an elevated level of HCG is still expected at 6 weeks post-D&C. I was scheduled for another blood test 4 weeks later. A couple days later I called the doctor and said that I wasn't comfortable waiting that long to see if the levels had risen and that I wanted to come in sooner. This was my first step toward not letting doctors dictate my course of action. My husband was tremendously supportive in this, reminding me that it was *my* body and that *I* needed to call the shots. I love him for that, because I think that was the most important lesson I learned during the experience: to take control of my life and my body.

I had another blood test at 2 weeks instead of 4, and my levels had dropped to 8. It was quickly going in the right direction. I wish that my HCG levels had been taken when I was pregnant so we'd have that data too, but with no history of problems, there really was no reason to do that type of workup.

I had follow-up blood tests in February, March, April, and May, with all the HCG results below 5. It was at the April visit that I asked more questions, including whether we needed to wait until June—the official 6-month mark since reaching a normal level. My doctor stated that although it was recommended, another month would be sufficient, because my levels so immediately declined and stayed below 5, my periods and ovulation returned immediately to a 28-day cycle in January, and my overall health was just fine. So when the doctor called after my May visit to confirm that my levels, as expected, were below 5, I was cleared! That meant we could try again to conceive. As relieved as I was, I now for the first time in began to think, "will it happen again? Of course it won't happen as quickly as last time, so am I ready for several months of trying?"

We didn't try in late May, but I knew which week in June would be the best to try, so we did. As my new turn of good luck would have it, the home pregnancy test that I took 4th of July weekend in 2006 was positive! I called the doctor on Monday, and I think the office staff was even happier than we were. We talked about my history and how best to proceed. They agreed to see me immediately to start bloodwork to track my HCG levels and make sure they were doubling. My first visit was July 7, and I had bloodwork to test HCG and progesterone each week in July. All looked okay until the last week, when my progesterone was 14. The doctor said the HCG level was great, but that they like to see progesterone levels above 16. I was immediately put on prometrium suppositories.

In my mind, it was happening again. We had told no one of the pregnancy this time. My husband and I just focused on building our house, which now had been going on for about a year, and just tried to get through each day. The morning sickness and some aversions to poultry and grains took hold as well. It was at the September 5, 2006, visit at 12 weeks that the doctor said that my progesterone levels from August had risen to 43 and that at this stage the placenta took over, so the prometrium was no longer needed. I finally told my parents about the pregnancy around this time; I had needed to clear the first trimester. The nuchal translucency was that week, the results of which were great. The alpha fetoprotein test at 16 weeks was also normal. At 16 weeks I started telling my family. By 20 weeks I was starting to show and after the next ultrasound I told people at my work about the pregnancy. It was real now. It felt scary and exciting. I didn't want to know if it was a boy or a girl; I wasn't ready to identify with the baby until it arrived and I knew it was healthy. I was still protecting myself, I guess. It wasn't until my baby shower in

February when I registered for all the baby stuff that I truly embraced the baby and the pregnancy, and it was great to finally feel so healthy and happy.

On St. Patrick's Day 2007, 4 days past my due date, I didn't feel quite right and the cramping started. I told my husband that I thought he should grab something to eat because we were going to have a long night ahead of us. At 7:09 A.M. the following morning, after 3 hours of pushing (whew!), Joshua was born. He was healthy and beautiful at 7 lb, 7 oz. I breathed the biggest sigh of relief in my life.

I'm one of the positive stories that I found so hard to find when I was struggling to cope with my loss, and I hope my story gives hope to others learning of their partial molar pregnancies.

Julia

We got pregnant at the end of January, 2005. We have a little boy, Johnny, who was turning 1 that February, and we had had some fertility problems with him, so we were excited that we had gotten pregnant on our own this time. Things were fine at the beginning. I was growing big very quickly, and by 12 weeks was starting to switch to some maternity clothes, which seemed early to me. But of course I had had my son only a year earlier, and they say the second "pops out" much faster. I had quite a few ultrasounds because I was older and had had a miscarriage before my son. All signs pointed to a healthy baby with a strong heart.

I took my son to the park to meet friends in early April, and I felt like I was wetting my pants. After finding a stinky port-a-potty, I discovered I was bleeding quite heavily. I packed up my son, called my obstetrician, and headed for the closest office. My husband was on his way, and it seemed like I was forever in the waiting room as I sat on a towel. My son also opened a door for the first time while I was in the bathroom with him trying to clean myself up—not exactly a milestone to be proud of. We had an ultrasound, which showed a healthy baby, and we were put on a "wait and watch" approach with partial bed rest. The bleeding stopped, and I was feeling good.

Two weeks later I started spotting again. After another ultrasound, I was put on complete bed rest. This was rough because I was a stay-at-home mom. My husband tried to work less, and I thank God for my dad because he took Johnny every day. My mom was with her parents, because my grandpa was being put into a nursing home. It was a bad time all around. I was seeing a doctor once a week for ultrasounds. I did get a thrill when I felt her kicking; I had felt my son kicking early also. The bleeding became less red and more of the brown prune juice color. We had heard from doctors and nurses that brown is better, it means it is old and not fresh. At this point I started to see a specialist in high-risk births while also seeing my obstetrician.

At 18 weeks I had a triple screen test for abnormalities. It came back with several bad indicators. I was immediately scheduled for an amniocentesis at my next ultrasound. The technician didn't say a word about the ultrasound. We waited for the specialist, who examined me again, and then we went into his office. "The baby has passed sometime in the last week." (Slap in the face.) "You also have a partial molar pregnancy, which is cancer." (Slap in the face #2.) My son was very cranky, so my husband had to take him into the lobby while I listened to the details of having a D&E. Although I was far along, inducing birth would cause too much bleeding. Among those details were the facts about cancer and the conclusion that my baby would never have survived. I remember distinctly not crying hysterically until we got my son home and into bed for a long awaited nap.

The next day I went into downtown Detroit to have seaweed pushed into my cervix to help soften everything for the surgery the next day. It was actually painful, and I did not have any medication, not even ibuprofen. The next day was the surgery. I was put completely under and lost a liter of blood. I awoke with a bad headache and would not stop vomiting, but I convinced them I was fine so that I could go home. The next morning I still had a horrible headache. Eventually I had to have a spinal blood patch to fix a "spinal headache." During that time I glanced at my chart and saw my HCG level was 1.4 million.

A week later I felt emotionally and physically sick. I kept having bad abdominal cramps and only felt relief while laying on my side. After another ultrasound we discovered my ovaries were being overstimulated by the hormones and were causing the cramps. This was a real low point for me, because I really wanted to get back to my stay-at-home mom duties and stop feeling so sick. I also watched as many pregnant friends grew larger and more glowing every day.

It took a long time for my HCG levels to drop to zero, because they started so high. In a way I felt a little relieved that I had to wait 5 months, because the thought of trying again was overwhelming. During this time I passed my due date in early October. I watched as three of my friends had babies in that same month. I was the odd one out, and I was still on cancer watch. I found some support online at the MyMolarPregnancy Web site and also in a support group at my local hospital, where there was a group called HUGS that was specifically for baby loss. There are not many groups like it, and obviously there were no other women with molar pregnancies there, but it was a great support for me. If there is any chance for you to join a face-to-face group like this I recommend it highly.

The 6-month wait after my HCG level hit normal was awful. I was so ready to try again. Once we were cleared we got pregnant again right away, but it resulted in an early miscarriage. The D&C for this second miscarriage took place on the exact same day as the D&E the year before: May 5. That date is now approaching again, and I am still not pregnant in spite of many months of Clomid treatment and fertility monitors. We are not young parents, and at some point I suppose we will just have to stop for our own sanity. My husband doesn't want to change diapers when he is 50, and I don't blame him. I am 8 years younger. We are trying more aggressive testing and will try for 6 more months or so before giving up. It is so sad to think I went through all of this and may still not end up with a baby.

Both of the babies I lost were girls. I really, really want a girl and always have. I feel like I am being teased. I hope in the next year I have that baby we want so badly or find the peace to know that we are done.

Julie

In March 1993 I became pregnant with our fourth child. We had had three children in just under 4 years, and I wanted one more baby before calling it a day. I had not had any problems conceiving the first three babies, but this time it took 17 months; the previous pregnancy had occurred while I was still on the pill, so I thought it was strange to have to try so hard for another.

Eventually, on April 5, I had a STRONGLY positive pregnancy test. I was thrilled! The baby was due at the beginning of December. I went to a little shop near our house and bought a tiny pair of socks to show my husband when he came home. We were happy that it had happened at last!

Yet my initial elation was short-lived; I began to feel that something was wrong, and I told my husband so. He said I was probably just nervous because it had taken so long, and I hadn't been pregnant for over 3 years. Others to whom I fretted said the same. We decided we needed a break—just one night away, but alone, so we could talk and be together. My friend, who had just learned I was pregnant, said she would look after the kids for us. On Thursday, May 13, I went to visit her. She congratulated me, but I told her that I felt unwell and was worried. I kept getting "pulling" pains in my lower abdomen and just felt odd. I said that I thought something horrible was going to happen. That night, it did.

The kids were in bed, and I was ironing the things we were taking away with us when I got a cramp and began bleeding. I screamed out to my husband. He said he'd never heard such an awful scream before. He was saying, "What should I do? I don't know what to do." I told him to get the doctor. I sobbed, I was just so shocked. I had never had any real problems before and never dreamed that miscarriage would happen to me.

The doctor came, examined me, and said he thought it was twins, but that I was miscarrying one or both of them. He thought this because my uterus was big for the date. Because I have a family history of twins, this seemed a likely explanation to me. He said to call him back in the morning or sooner if the bleeding got worse. I called my friend, and she said she would take the children anyway the next morning so that we could concentrate on what was happening without worrying about them.

I couldn't sleep that night—I was worried sick, the cramping was worse, and I was losing lots of blood. My friend came for the kids at about 8:00 A.M. the next morning, and I will always remember the look of sorrow on her face. She knew how much this baby was wanted, and she too was shocked that this was happening. By 9:00 A.M. I could not stand anymore, so we called the doctor back. He reexamined me and again said that he thought it was twins. He said the cervix was closed, which was a good sign, so not to lose hope. He then sent me to the hospital. In the taxi on the way there, we were joking that it was twins and that they must be having a fight in there. I thought, "I'll need another double buggy!" (I'd had one for the boys) and was imagining life with five children!

I remember later lying on a gurney and a doctor coming to put an IV cannula in my hand, because I was bleeding quite heavily. He was jabbing at my hand trying to get the darn thing in, blissfully unaware of the pain it was causing (and the blood all over the sheet), then he finally looked up at me and said, "So, you're bleeding vaginally—do you think you could be pregnant?" I couldn't believe it! He hadn't asked anyone what was going on, nor read my chart. If I hadn't felt so weak, worried, and ill, I'd have told him what I thought of him. The nurse who came in later was annoyed at his remark and the way he had inserted the cannula, and she was very sympathetic.

I finally went to the gynecology ward at about 1:00 P.M. A doctor examined me and confirmed that it could possibly be twins and that the cervix was still closed. He patted my hand and said, "That's a good sign, don't give up hope." At this stage, I still believed him. I just lay there all day in pain and worry about the kids and my husband but trying to hold on and believe that the bleeding would stop and I would soon be back home again. That was on Friday.

All day Saturday I bled and bled. I was examined by a couple of nurses and then a doctor and was again told my cervix was still closed, so not to lose hope. I now wondered who they were trying to kid—no baby could survive that much blood loss. My mum had come to visit, and she was horrified at the amount I was losing; I was literally lying in a pool of blood. I had gone through a few sheets already, and kept telling my husband to get the nurse because I thought it needed changing again. When she came, even she was shocked by the loss, but she quickly recovered herself and said, "Oh, I've seen worse, don't worry." I thought, "Where have you seen worse, then—in the morgue?" I know she was trying to help, but I didn't want false hope. I could see for myself what was happening. Also, the pain was really bad now; it felt like a mini labor. I was given pethidine for it and then collapsed. The nurse had been gone about 5 minutes after administering the drug when I felt like I was sinking into blackness. I said I felt strange, then I knew no more. My husband pressed the emergency button, bringing four nurses running. They called a doctor. When I came to again, I was on oxygen. I had had a reaction to the pethidine, which dropped my blood pressure and caused the collapse.

My tummy was growing by the minute; it was now the size of a 20-week pregnancy. The doctor eventually got a portable scanner to see what was happening. I saw a fetus right in the middle of the screen; it looked about 7 weeks (I was nearly 12 by now) and had no heartbeat. I asked if there was a heartbeat, but the doctor wouldn't commit herself; she said because it was a portable scanner she couldn't see clearly enough. She thought I might have fibroids. My sister-in-law had arrived by this time; she couldn't look at me while the scan was happening because she too had seen what I saw. I was booked for an emergency scan the next day, Sunday. (The radiographer had to be called in out-of-hours for this.) I passed the rest of that night in a dream-like state; my mind was almost numb. I went through the motions of having my sheets changed, using the toilet, and talking to the nurses, but I felt detached. I wanted to sleep, curl up in a ball and be left alone with my baby for as long as it was still in there. I hardly even thought of my kids at home. I knew they were okay, but this little one wasn't, and I felt sick with guilt and sorrow.

The next day dawned, and the dreaded scan was upon me. I knew it was going to be bad news, and I just didn't want to hear it. I felt so physically weak by now I could barely sit up in the wheelchair as I was taken to the scan room, and every time I moved I lost more blood.

All the radiographer said was, "Have you been very sick with this pregnancy?" I told her I had felt really nauseous but hadn't actually vomited, which was a first for me. I knew from her face that something was badly wrong, but I'd never have dreamed what it actually was. I was told when I got back to the ward. When the nurse said the words "molar pregnancy" I said, "Oh, you mean a hydatidiform mole?" She was amazed that I'd heard of it, but I had read about one in a magazine just a week or so before. I thought at the time what a horrible thing it was. As I lay there trying to take it in, my mind seemed to shut down at first, then the sobbing started. I wanted to die; I just didn't want to hear that my baby was gone. The nurse who was looking after me was lovely; she was so kind and was almost in tears herself. She got me a private room free of charge. She and her colleagues were so caring, and I will remember them always.

The next day I had to have the pregnancy "removed." The consultant who was put in charge of my care had only been at the hospital for 3 months, and this was his first experience of molar pregnancy. He told me he would do his very best to avoid it but that I might need a hysterectomy because of the size of the uterus (it was now 24–26 week size) and the blood loss. I was in surgery for almost 3 hours, and when I woke up I was so relieved to still have a womb. The physical pain was nothing compared with the mental pain I felt at having lost my baby. That's what it was to me—my baby. It meant nothing to me that the pregnancy was all wrong and there was no chance of a viable fetus. It was still my child, and I still grieved.

Later that day (or it may have been the next day, I'm not sure), a nurse came to my room to explain some things. She was really nice and kind to us, but I was taken aback when she started talking about the "prognosis" and "outlook." I thought to myself, "Good grief, she's talking 'cancer-speak'!" She told us about the follow-up I would need and that I couldn't try for another baby for 2 years. That was the final nail in the coffin. I was absolutely distraught. We were also told that the "conceptus" was a male karyotype. We named him Adam Joseph.

At some point during my time in hospital, I was sent for a chest x-ray to look for secondary tumors. During the x-ray I was thinking over and over, "I'm 29 years old and was having a baby; how come I'm now being tested to see if I have cancer? What if I die, what will happen to my husband and the kids?" To say that my mind was in turmoil is a monumental understatement.

My HCG levels had been way over 100,000 before surgery, and when I saw yet another consultant he said they were now coming down nicely. He told me bits about why the levels needed to be monitored, and kindly informed me that if left untreated, any residual molar tissue could travel to the brain via the bloodstream and be rapidly fatal, thus the monitoring was necessary. Now, I'm the kind of person who wants to know everything that's happening with my body, but that felt like a punch in the stomach. I was reeling. Thank God for the nurses, who were a bit more thoughtful and put things into perspective. I don't think the consultant intended to upset me, he just didn't seem to realize the import of his words. I was also anemic and was started on iron therapy.

When my husband brought the kids in to see me, my heart broke for them: they were being so good but looked so lost. My husband told me that when he got them together and told them what had happened, they got in a little huddle with their arms around each other and sobbed their hearts out. They were very young: Zoë was 7, Daniel was 5, and Stephen was

3, and it was hard for them to comprehend what was going on. I remember Daniel drawing a picture at school of me in my hospital bed, complete with drips and sad face. When he gave it to me, I had to hobble to the bathroom to have a good cry: it broke my heart. I heard him saying to himself as he sat in the corner with big 5-year-old eyes looking at me, "I've seen my mummy now, and I'm fine."

Zoë wrote a letter saying she was sorry that our baby had died, and I later learned that she felt guilty for being naughty when I was pregnant and was worried that she may have caused what happened. My poor baby girl! My little Stevie was only 3 and didn't understand much except that there was no baby anymore and that we were all sad. He was so well behaved for his daddy and grandma, even though he was missing me so much.

I went through so many emotions at this time, and to try to make sense of it I kept a journal that I titled "Catalogue of Disaster." I still cry now if I read it, even after 13 years. I was hurt, bewildered, guilty, scared, and very, very angry. I felt guilty for bringing all this on my family. I was blindingly angry when people told me to be happy with the children I already had and not to be selfish in wanting another. I hated people saying that I was "only" 3 months and at least it wasn't a stillbirth. I hated telling people it was an ordinary miscarriage to avoid all the dumb questions and all the explanations. I hated that the focus was on me and my health; I had lost my baby, for Pete's sake, was it too much to ask for people to remember that? I didn't care that there was a problem right from conception and that he wouldn't have survived anyway. He was *my* baby and I wanted him back. I didn't care that my own health was in jeopardy. I WANTED MY BABY! I didn't want flowers, they made it feel like a funeral. I didn't want people to say things intended to make me feel better—I wanted them to acknowledge that I had been pregnant and just cry with me or at least let me cry.

For months after I kept waking up thinking I could hear my baby crying. I kept feeling as though I'd lost something somewhere and couldn't find it. My heart would pound and I would shake and have this feeling of dread. My arms ached to hold my baby. I also remember feeling like a freak; no one had heard of molar pregnancy, and if I tried to explain, it sounded so weird. It's no wonder most people just didn't "get it." As soon as the possibility of cancer was mentioned, well, that *really* threw them and they just didn't know what to say. Some people actually avoided me so as not to talk about it. I know it must have been hard for family and friends, but this was one of those situations where I found myself wishing more than once that people would think before they spoke. My good friend Debbie helped to keep me sane, however, and even managed to make me laugh.

I wanted to go back and do this pregnancy over again, only this time he would be fine and all would be well. Better still, I wanted to wake up and find it had all been a nightmare. I left the hospital after 8 days, but I was back in a few weeks later with heavy bleeding. My general practitioner sent me in for another D&C, but the bumbling doctor on the ward (who I hadn't seen before) decided to wait and see if the bleeding stopped by itself. He sent bloodwork off to a specialist lab in Liverpool and expected me to wait on the ward for a week for the results. I wasn't happy with this; I didn't want to leave the kids again for any longer than necessary. One night I walked off the ward—still in my nightgown—intending to go home because a nurse had told me not to get upset, and my husband had been moaning about something or other that seemed petty when I was still so ill. I just wanted to scream my lungs out.

I saw a number of doctors and most of them seemed a bit unsure of what to do with me. This added further to my feeling like a freak; if the medics didn't understand much about this condition, how could anyone else? After 3 days I was sent home with drugs to stop the bleeding, but it didn't stop; I eventually hemorrhaged and was admitted again. I wryly joked with the shocked nurse that she should issue me a season ticket and my own suite of rooms.

This time I saw the consultant who did the original "evacuation," and I was so relieved; he was very sympathetic and held my hand and said, "Don't worry, I'll sort you out." He examined me and removed lots of blood clots off the cervix. (Painful!) He ordered a D&C because he thought I had an invasive mole. I was so depressed at this point I didn't really care if it was. My thought processes were a bit muddled, to say the least. Thank God it *wasn't* an invasive mole; it was just "debris" and blood clots. The D&C seemed to help, and the blood loss slowed to manageable levels. Once again I was given a private room, and once again the nursing staff were very good, but I wished the D&C had been done the on the previous admission because that would have saved a lot of upset! I returned home to the rounds of blood draws and urine collection. I had to collect urine for 24 hours then send a sample to the Jessop Women's Hospital in Sheffield once a week for 3 months, then fortnightly for another month, then once a month until the levels were normal. I think it took 6 or 7 months to reach that stage. Then the tests were cut back to quarterly.

The worry during this time, quite apart from whether the levels were dropping, was how we would get to Sheffield for treatment if they did not drop. We had no car, we had three young children, and my husband worked three different shifts including nights. Had I needed treatment at Jessop I would either have had to go and stay as an inpatient or face a lot of traveling. As each blood and urine test came back on target, I felt relieved that I didn't need further treatment and all the attendant inconvenience and disruption—not to mention worry. My heart bled for the women I knew or had heard about who weren't so fortunate.

I had been told by the hospital consultant and by Jessop to wait for 2 years after I had reached normal levels before trying to conceive again. I couldn't use the pill because it gives false readings, so we had to go back to using condoms. Not that I wanted very much to make love anyway; I still felt pregnant and sick for quite a while, and I absolutely hated having to use contraception when all I wanted was another baby. I understood the reasons why I couldn't try yet, but every month my period made me weep. I remember thinking too how cruel it was that if I had taken pregnancy tests during this time, they would have come back positive.

Another cruel irony involved our house; when I found I was pregnant, we decided we would extend it to have an extra bedroom. After the diagnosis, we thought we may as well go ahead and have it anyway, with an eye to the future and a successful pregnancy at some point. We got planning permission, and I started to look forward to having it done. It was something to focus on, a distraction. Then one day when I was feeling particularly weepy my husband called me in from the garden. He told me he had just been to the Building Society about life insurance in regard to the extension on the house. They'd refused to insure me because of my condition and had told him to come back in 6 months, at which time they would review my case. I went berserk. I was convinced that I was dying and everyone was keeping it from me. I was totally illogical, but at the time I couldn't see that. I thought the doctors had told my husband I was terminal but not to tell me and that he had gone along with it to protect me. I thought they had given me false test results. I threatened him with all sorts of things to get him to tell me the "truth" and said I'd never forgive him if I found out he knew I was dying and hadn't told me. I called my mum, screaming and crying and asking what I'd done to deserve this. I railed against God — something I'd never, ever done — and told Him to leave me alone, I didn't want to know Him anymore. I thought I was losing my mind. It was just horrible.

My husband eventually persuaded me that I had got the wrong end of the stick, and I calmed down. From that day to this, though, I have a horror of ever being in that situation, and he has promised that he would never keep any such information from me to try to "protect" me. Some people would rather not know if they were dying, but I am not one of them (and we eventually got our insurance elsewhere!).

When my HCG levels were back to normal, I started itching to try and conceive again. We were given odds of 1 in 75 for a recurrence of the molar. I thought, "Well, I have 74 chances that it will be okay!" (Whether those are safe odds or not, I don't know. Math has never been my strong point!) We held off trying until 15 months from diagnosis. This was sooner than I had been told, but I figured it may take a while anyway, and I didn't want to wait any longer. This time, it took 18 months to conceive.

Three years and 3 months after my molar pregnancy, in August 1996, I gave birth to Ingrid by emergency caesarean. As soon as the pregnancy was confirmed I was given an urgent appointment with the consultant who had looked after me before. I was 6 weeks pregnant and petrified to go to see him. I had to go alone because my husband was looking after the kids, and I was surprised when the consultant opted to do a scan there and then. I was so scared, and I lay there with my eyes tightly shut. I felt like I was a time bomb that could detonate at any minute with the dreaded news, "Sorry, it's another molar." But he eventually said, "Open your eyes, it's fine. There's the heartbeat!" I looked and there it was, a little light flicking on and off on the screen. The relief was just indescribable. He said to me, "As far as I'm concerned, this is a completely normal pregnancy. Go home and enjoy it." I wanted to kiss him!

I then had scans at 9, 12, 15, 23, and 32 weeks. I wouldn't allow myself to get excited this time, though, and wouldn't buy things for the baby until I was about 5 or 6 months. I had a few scares along the way, including possible premature labor, and was worried that something awful would happen. I also knew that another pregnancy—even a normal one—could cause a recurrence of the mole, and I knew that I had to send a urine sample to Sheffield 6 weeks postpartum to check things were still normal, so that was in the back of my mind the whole time. I tried not to think too much about it, but it was there all the same. Once you have a pregnancy go wrong, that innocence you had before is gone forever. I was one of those women who believed miscarriage (of any type) would never happen to me. How naïve I was!

The pregnancy wasn't easy—physically, emotionally, every which way—and I was terrified the whole time I carried her, but she was worth it, and I'm glad I ignored the people who tried to put me off getting pregnant again. They meant well, I know, but I had to try one last time and end on a triumphant note. I'm so grateful that I did, and my message to others who have had or are going through this nightmare is please, never lose hope. You CAN have a successful pregnancy after a mole. I know how hard and scary it is to believe that next time it will be okay and you will get your longed-for baby, but it is possible. Ingrid is living proof of that. Also, remember that you're not alone. It's wonderful that there is so much support and information out there now, through sites like MyMolarPregnancy; when I lost Adam, there was nothing, and I felt so isolated.

As I said, Ingrid was born by emergency cesarean—within minutes of her life—and was in the neonatal unit for her first few days. Upon waking from the anesthetic I was told that she was shocked and traumatized and was in special care, and that they thought there may be brain damage. We were both ill for the first weeks and months of her life, and when she was about 4 months old, the molar pregnancy reared its head again with a nasty little surprise. I had been in almost constant pain since her birth and could hardly hold myself upright when walking. One day when I was at the clinic with Ingrid for one of her vaccinations I told the doctor I could not take anymore. He told me—rather brusquely—to book an appointment, so I did and went back a few days later. I explained what was happening and he said he wanted to examine me. He palpated my tummy, and his face paled. He said, "I think there may be a recurrence of molar tissue. I'm booking an urgent appointment with Mr. —." All I could think at first was, "Serves you right for being funny with me the other day!" Again, I wasn't thinking straight.

My husband was shocked when I told him—to say the least—and again I felt guilty, as though I was a nuisance. I thought, "Surely I can't be going through it all again. I have the baby now, too!" My mind was filled with what-ifs and how-will-I's, and I was running on autopilot.

I went to see the consultant 2 days later: thank God, it wasn't a recurrence. He told me my uterus was too bulky because it had not contracted properly after the c-section, and he suspected adhesions. He did a laparoscopy a few weeks later, after which he recommended a hysterectomy: it was a mess in there, my periods were horrendous, and I was anemic. So when Ingrid was 16 months old, I had a hysterectomy. I didn't care about losing my womb; I had my baby now, and our family was complete. I also wanted the pain to stop. Best of all, I knew that once the uterus was gone, I would never have to worry about a recurrence again. After a long recovery I started to feel better and was thrilled that my iron level went up from around 9 to 12.8!

I knew—and had known for a long time—that things with Ingrid weren't so good, however, and when she was 19 months old I got my Health Visitor (a nurse or midwife in the United Kingdom who helps new families) out to tell her my concerns. Ingrid was referred to a pediatric consultant and was diagnosed with cerebral palsy at the age of 20 months.

She is a wonderful child and was worth all the heartache I went through to finally get her. Despite everything, I am so blessed to have her, and I thank the Lord that having the molar pregnancy didn't put me off trying again. I often think that had everything been normal with Adam, I'd never have had Ingrid, which is a strange thought. We have to believe that everything happens for a reason, and as a Christian, I trust that God is in control—no matter how things seem. I will always think of myself as a mother of five, and I look forward to seeing my little lost baby one day, where he will be perfect and whole. 'Til then, baby…

I hope that reading my story—our story—will be of help to someone, and I pray God's strength and help for all women who are living through this. I wish you healthy, happy pregnancies and beautiful babies in the future.

Karen

I had a molar pregnancy in September 2001. I was only just turning 24 then, and I was working away from home at a holiday site. At the time I had only heard of pregnancies or miscarriages; I had never heard of a molar pregnancy. I found out that I was pregnant and was shocked. I told my boyfriend and my parents, and they were all over the moon. We decided that we would carry on working at the holiday site until after Christmas so that we had some money, and then we would move back into my parents' house. However, at the beginning of November, I became really ill. I couldn't stop being sick and couldn't keep anything down. I knew I had to give up my job and go home. My boyfriend, Jamie, and I finished work and headed back to my parents' home in Cornwall. I had already started spotting by then and didn't have any energy. It was the worst time of my life.

When we arrived home I had an appointment with the doctor. He said all was fine, and the midwife came a few days later and listened to what she thought was the baby's heartbeat. Within a few more days I couldn't even open my hands. My boyfriend and my parents rushed me back to the doctor, and he looked me over again. He could see that I wasn't well, so I had to do a water test, and when he tested it he found traces of protein, then phoned the hospital and booked me in. When I got there they put me on an IV drip and made me blow into a paper bag, which I couldn't do because my hands wouldn't open. The next day I went for an ultrasound. I was so hoping that everything was going to be fine, but the woman had a look, and all she could say to us was that there was nothing there. Then she put us into the waiting room. (Who was she to tell us that we had no baby? She was not a doctor.)

We returned to the ward, where the midwife said that the doctor would be up soon to speak to me. None of us could understand what was going on. The doctor came in and informed us that I had had a molar pregnancy, a complete molar pregnancy, so there was no baby, and that it was a 1 in 1,200 chance of having it. I had my D&C in the next few days but lost two pints of blood, so I had to stay in longer than I had hoped. I returned home just in time for Christmas.

Just after Christmas I received a letter instructing me to have another ultrasound to check that my womb was all clear. The doctor did this ultrasound, and he told me that in 1 in 10 cases the complete mole keeps growing and could burrow into the lining of the womb and possibly spread to other organs—a persistent or invasive mole. He then said that it looked like mine was heading that way, turning into what they call choriocarcinoma, which was cancerous. He referred me to Charing Cross Hospital in London, and they asked if I could come in a couple of days to begin treatment. My mum came with me, because it was a long way from Cornwall. I had a week of the low-risk treatment at the hospital, then returned home and had to go to the doctor's office for the rest of my treatment.

At the time, my Nan was dying from cancer in her gullet, and around the same time she died I was told that my chemo was not bringing my HCG level down and to stop treatment until they decided what to do next. Two days later my mum and I were back in London for the start of the new treatment. I stayed another week and had the higher-risk treatment to make my levels go down faster. I also had to have three lumbar punctures, one every other week, because the cells had spread to one of my lungs. I returned home to finish my treatment at the local hospital. All in all, I would say that I visited Charing Cross Hospital fewer than 10 times, and I am happy to report that I have been given the all-clear—my cancer is gone—so in the future if I become pregnant I just hope that it does not return again.

Katie

My name is Katie, and I am now 28 years old. My husband and I were married in February 2002. I went off of birth control in October 2002 after having used birth control pills continuously for 6 years. We were able to get pregnant the first month we tried, in January 2003. At the time, I was 25 years old, and I had a "normal" pregnancy. I was nauseated but only vomited once during the first trimester. I had typical symptoms, and the bloodwork they took at 9 weeks came back normal. We went in and heard the heartbeat at 13 and 16 weeks, it was a strong heartbeat. When I was 17 weeks, I became violently ill and went to the emergency room. The pain was in my stomach, my upper abdomen, and after consulting with the emergency-room doctors and my obstetrician it was decided that I either had food poisoning or a stomach virus. I was in such great pain, but it only lasted one night. We now believe this might have been when our baby died, but that is pure conjecture. I never had spotting of any kind throughout my pregnancy, and it could be that I did, in fact, have food poisoning entirely unrelated to my molar pregnancy. It did not worry me at the time, because I innocently assumed that if something were wrong with the baby, or if the baby had passed, I would have cramping and bleeding. But I was just "fine" the next day.

My husband and I arrived at the obstetrician's office the day before my 21st week of pregnancy for our ultrasound. Things were immediately wrong. The baby was too small, and no heartbeat could be detected. My placenta was huge and had spots. We were in shock. We were referred across the street to the hospital for a second ultrasound. This confirmed that there was no heartbeat. I still have an image in my head of what our precious baby looked like, so adorable with its hands clutched together. I am grateful for this memory, despite the sadness of the image. We know our baby survived at least through 16 weeks but was about the size of a 13 week old. I am grateful that my husband was there with me.

I was given a choice of being induced immediately or the next day. I was terrified. I chose the next day. My husband and I went home and cried together all night and read all we could about partial molar pregnancies on the Internet. I woke early the next morning, and we got to the hospital by 6 A.M. By that time the doctor had consulted with her colleagues and determined that I was at risk of hemorrhaging if I was induced, so we decided on a D&E instead. I was relieved that I would not be awake for the procedure. My blood pressure was 154/90—it is normally 110/90 or lower—and my HCG was over 200,000, yet I had arrived at the doctor the previous day as normal as could be. I had not had spotting or cramps or discomfort other than the one night in the emergency room. I was sent for a chest x-ray, which came out okay.

I was given anesthesia and sent in for the D&E. It took longer than expected, about 75 minutes. My obstetrician said even she was shocked by the volume of blood I lost; I almost had to have a transfusion. As it turns out, I was already dilated when they began, and I would likely have miscarried soon. I stayed overnight in the hospital. The next day I was discharged, and I was lucky in that I needed only ibuprofen, not narcotics, for the pain. I was very anemic but was given iron. My husband and I grieved with family and friends, and we planted a tree in our backyard to remember our baby. I also purchased a silver ring to wear, by which I would remember our baby. I had hoped to find out the sex of our baby when the reports come in. When got the results of the chromosomal tests back, they showed XX. Mostly likely the sample was contaminated with my own tissue, but several other scenarios are possible. I will probably never know. My HCG levels were over 200,000 before surgery and down to 50,000 afterward. One week later they were at 3,200. From there they steadily lowered and hit zero 8 weeks after my surgery, about the end of July 2003. My doctor initially suggested that we wait 6 months from this point to try to conceive again.

I am grateful that we found out when we did. If I had had an ultrasound earlier, or if I had elected the alpha fetoprotein screening, we would have found out when the baby was still alive and would have had to wait for the baby to pass away, which might have been more difficult, or I may have been advised to have surgery while the baby was still alive. However, if I had not gone to the doctor when I did, I probably would have miscarried naturally and lost a lot of blood, and my health would have been at risk. It took a lot to work through our grief; I really appreciated the fact that my obstetrician practiced on her own, not in a huge group, so that everyone at her office knew me and my circumstances, and they were so kind and patient with me. I am happy to have my health, but I cannot describe how much my husband and I loved our baby. We know our baby is watching and loving us from heaven.

After my HCG level reached zero, I was monitored on a monthly basis. My HCG never increased, and my doctor cleared us to start trying to conceive in early December of 2003, just 5 months after I hit zero. She felt this would be safe and was very understanding of our desire to try to conceive again. I chose not to go on the birth control pill during the 5-

month period, instead relying on other methods of birth control. I remain wary about the link between my partial molar pregnancy and the fact that I had been on birth control pills for so long, and so close to getting pregnant. I don't know if there are any studies establishing such a link, but I did what made me most comfortable, and what made me worry the least—I figured that was what was best for my health, even if there was no science behind it.

We got the all-clear to conceive again in December, and I got pregnant again in January. I was wary throughout the first half of my second pregnancy, but it was successful! I was monitored by my doctor not with bloodwork but rather through a series of ultrasounds during my first trimester and even into the second trimester. She was wonderful and supportive and understood my fear and panic about the situation. My son Tommy was born at 39 weeks and weighed 8 lbs 2 oz. My placenta was tested after my delivery and checked out just fine. Tommy is doing wonderfully and will turn 2 years old in a week. I once again chose not to go on the pill for birth control after Tommy was born. It may be paranoia, but it makes me feel more at ease. I am also a vegetarian, but I tried to eat some turkey from time to time while we were trying to conceive, although I really am not convinced that this had anything to do with my problems, because I eat a very healthy and nutrient-packed diet, take vitamins, and so on. I am now once again 16 weeks pregnant, and I didn't change my diet at all prior to this pregnancy.

As with past pregnancies, we have been blessed in not having difficulty in conceiving. I am still nervous about a molar pregnancy, although I probably shouldn't be. This time, I had ultrasounds at 6 and 9 weeks, but other than that my pregnancy has been monitored in a "normal" fashion. Because of my first experience, until I feel unequivocal kicks,

I will not be at ease. I'm still not sure what caused my partial molar pregnancy, or why I had so few symptoms. I always thought that if something were wrong with the baby I would have cramping or spotting. Now that certainty has been taken away, and I don't think I'll ever get it back. In a way I am very scared, but in another way I feel prepared for anything after what I went through.

I still wear the ring to remember my first little baby, and when my husband and I moved, we purchased a new crape myrtle tree to plant in our yard, again to remember our first little one. We are thankful that we have been able to conceive so easily, and although this experience was heartbreaking, I believe it strengthened my faith and resolve as well as the relationship between me and my husband. It also made me truly appreciate the miracle of childbirth.

Kellie

My name is Kellie. I am 23 years old. I have been married for 3 years to my wonderful husband, Jim. We were married in October 2003, and in March 2004 we learned we were pregnant! I had a completely normal pregnancy, except for a few minor things, and I had a little boy, Joey, on November 3, 2004. Everything was great! We wanted to expand our family, so we kept trying, and in September 2005 we found out we were expecting our second child. I was so excited! Another little baby!

I would say that from the beginning things with this pregnancy were different. I am a very impatient person, so the day I was supposed to start my period I took a test, and it was a very faint positive. I called my obstetrician's office, and they said that I could have a blood test to confirm. I remember it was September 21. The results came back low, the number was 88, and they wanted me to have another test in a week. I had no idea what "low" meant, so I asked, and the nurse said that if my level did not go up significantly that I might be having a miscarriage. I couldn't believe it. I was so nervous the whole weekend. My first pregnancy was fine, why would I have a miscarriage? I had the second blood test, and it was in the normal range at 1,859. I was so relieved! I went to my first appointment on October 10. It's funny how I can remember all these dates yet forget what I just went into the kitchen for!

The appointment went well, and everything looked fine. The nurse said that my uterus was quite large; she even tried to hear the heartbeat. I am a fraternal twin and come from a family of twins in every generation, so my chances of having twins myself are pretty high. The nurse thought I should have an early ultrasound just to see. The ultrasound was set for November 1, and it showed one baby! We were very excited to see the baby for the first time. At that time I was about 10½ weeks along. During the ultrasound the due date kept popping up as June 7, but my original due date had been May 28. The ultrasound technician said she was not sure about my dates, so I would have to talk to my doctor. I didn't hear anything, so I assumed that everything was the same. Then, about a week before my next visit, the nurse called to say that my due date had indeed changed 10 days. I was freaking out. How could that be? I kept thinking about it, and it just didn't make sense. But we saw the heartbeat, and it was in the 170s, so I told myself everything was fine.

I was supposed to have my 12-week appointment (based on the new date) on December 4. But on November 29, our lives forever changed. It was the week after Thanksgiving. My best friend, Kate, and I went to the mall to do some Christmas shopping. The night before I had woken up with pain in my upper stomach. It was actually so bad that I had thought if I checked I would be spotting. I didn't really sleep all that well the rest of that night, but when I got out of bed for the day I felt much better, so Kate and I headed to the mall as planned. I had told her about the night before and also told her that a few nights before I had had a dream that I gave birth to my child, but it wasn't alive. She said that the dream wasn't a big deal, but I should call the doctor if I got the pain again. While we were shopping the pain came back, but worse. I couldn't stand up straight, and I couldn't even push my son in his stroller. I sat on a bench and called the doctor, and they wanted me to come right in. They thought I might have a bladder infection or maybe even something to do with my gallbladder. I sent Joey off with Kate and headed to the doctor. I remember kissing Joey goodbye and thinking to myself that things seemed different. I was so nervous. I went to the doctor, and my urine test came back negative for a bladder infection. They didn't know what was wrong, so next the doctor tried to hear the heartbeat. For 20 minutes there was nothing. She would try to hear it, then try to feel where my uterus was. She did this about four times, and would just say, "well maybe your uterus is laying low." She could see that I was very worried. It was too late for an ultrasound, so she sent me to the emergency room.

I just started crying, I was so scared. I went home and called Jim. He left work early and we went to the hospital. Even as all this was going on, I felt strange, but I still had it in the back of my head that everything would be fine. We got to the hospital about 5 P.M. We waited about 2 hours before they even called me back to an exam room. (That isn't so bad for an emergency room, but it was a very long 2 hours!) They wouldn't let Jim come back until I saw a doctor. The doctor came back and said they wanted me to have an ultrasound, then we would go from there. Again Jim was not allowed to come with me. The ultrasound technician was so sweet. I asked her before she started if I could have a picture—if everything was okay, that is. She said that they didn't do that. Right away I knew something was wrong. She kept asking me the first day of my last period. I just asked, "Is there a heartbeat?" She didn't answer me at first, then she said, "I'm sorry, no." All I remember after that was asking myself, how am I going to tell Jim? What did I do wrong? How could this be?

I had a D&C the next afternoon. I had not had any spotting. I didn't understand how I didn't know something was wrong. They said the baby had stopped growing at about 8 weeks. The next few days I was pretty much in shock. It wasn't real. Then on Friday, December 2, my doctor called. He said that the tests came back on the baby and I had had something called a partial molar pregnancy. WHAT? What is that? He explained everything to me, but I was so confused! I remember thinking, I am so upset about this loss, but now I have to worry about my health! My doctor said that I would have to get my blood taken every week until my numbers went down below 2. My first blood draw was a week and a few days after the D&C, and when it came back the level was at 98. At the time I had no idea if that was high or low. Later, I joined the MyMolarPregnancy support group and learned it was pretty low. Some women's numbers were in the hundred-thousands! It took 7 weeks for my level to fall below 2, then I had to have 3 more weeks of tests to make sure that it stayed that way. My doctors cleared us to conceive again in June. I went through so many emotions. I would cry all the time. But I know everything happens for a reason. I think the hardest thing for me was that my sister-in-law was pregnant, and we had been due 10 days apart. Her pregnancy was a constant reminder. I never really started to move on until she had her baby.

We have since found out we are expecting our second child (third in my mind). My due date is May 1, 2007. I am a complete nervous wreck and worry all time. I have come to find out more information about my molar pregnancy. With this third pregnancy I am considered high risk. At about 7 weeks I had some spotting. I went in right away for an ultrasound. To my relief everything looked fine, but when I was talking to the nurse about the results I told her how nervous I was. She told me, "well, we found a heart rate and once that happens, things normally progress fine." So I had to open my big mouth and say, "I had a heartbeat with the last baby." I was on hold for about 20 minutes and was then told that it looked like I had had two pregnancies. One had been a normal baby, and one had been the molar pregnancy. The mole then invaded the normal baby, which is supposedly an uncommon thing. I guess all I can do is wonder and hope that everything will go well with this baby.

I hope that everyone who reads my story will know that they are not alone. We are told that it is not common, but I think that all the women I have ever talked to on the subject feel very strongly about getting their stories out, getting the word out. Good luck to everyone.

Krista

Because I was older when I finally married (36 years), we decided to get pregnant immediately. In June 2001, I was pregnant and due in March 2002. We saw a heart flashing at the first ultrasound but found an overly enlarged yolk sac. I miscarried at 11 weeks, and genetic testing later identified a chromosomal error (on marker 14,15) of a female. Bad luck for me, I thought.

I have a short luteal phase, so the doctor tried us on Clomid and Crinone (progesterone), and within 3 months I was pregnant again. It all looked good, but there was never a heartbeat on the fetal pole, so we had to have another D&C at 9 weeks because my body didn't recognize the demise. Genetic testing revealed a girl with normal chromosomes but questioned whether it was my genetic material or the embryo. I became pregnant for the third time in June 2002 and was due the exact same week as my first pregnancy. This is where the story really begins.

I was feeling morning sickness all the time (unlike the other pregnancies). It was strange from the beginning, because I had some slight spotting at the time of what could have been implantation, and more spotting about the time my period would have started. I had never had anything like that with the other two pregnancies, and I just thought my body was having trouble recovering from the miscarriages and fertility medication. I went to see the obstetrician the following week, and she was going to put me on a high dose of progesterone, then have me stop in order to bring on my period. She first asked if I had taken a pregnancy test. Of course I had, but they were negative. She had me take one right then, and it came back a FAINT positive. She then ran a blood HCG test and called the next day to tell me that I was, in fact, pregnant. Because of my history, she wanted to see me weekly to monitor the progress.

At 6 weeks we saw a little deformed gestational sac (on vaginal ultrasound). At 7 weeks a fetus with a heart flashing was seen, but the measurements were slightly smaller than should be expected for my time of ovulation (I tracked it with a basal body temperature thermometer). At 8 weeks my husband came to the appointment, and we saw a heartbeat flashing again. I was finally bonding a little; maybe this one would work! At 9 weeks, no heartbeat was seen, but my HCG level was high. I was devastated, but I had been through this with the other miscarriages so I accepted it and had a D&C the next day, followed by a genetic workup. This time it was an XXY 69, which I learned later meant a partial molar pregnancy. My doctor placed me on birth control for the cycle after the D&C to help regulate my period more quickly. She also referred us to a genetic counselor. She did not mention a partial molar pregnancy, nor did she take the HCG level at the time. I don't think she was aware from the pathology that it WAS a partial molar pregnancy.

I was scared and did my usual thing to cope with the unknown: I researched it. I found that the XXY 69 was a partial molar pregnancy that occurs when an egg is fertilized by two sperm. I read many Web sites, some of which told me what I wanted to hear: that a partial molar pregnancy is RARELY cancerous. One medical site said they told their patients it was okay to work on another pregnancy as soon as they were comfortable with it, so that is what I did. When I found I was pregnant again just two cycles after the D&C,

however, I was shocked and worried that maybe this wasn't so smart. I had an HCG level (quantitative) run, which showed my level as 259, or 4.5 weeks pregnant, and just about right for me. I made an appointment to see my doctor. The first available appointment was when I was 8 weeks along. It showed an empty sac ("blighted ovum"), and another D&C was scheduled.

I decided to take a break after that. My body had been through so much. At this point we went to a university genetics clinic, and they studied all the genetic results (and considered us as carriers), but there was no pattern to any of the miscarriages, and the testing showed my husband and I had normal chromosomes. I became pregnant again a while later, but although we found a fetal pole at 9 weeks, there was no heartbeat. After yet another D&C the genetic testing revealed another random error in chromosomes.

This time, we were sent to infertility specialists. They said that none of the treatments available (e.g., in vitro, intrauterine insemination) could improve our odds. We could look into experimental PDG (preimplantation genetic diagnosis) in vitro fertilization or look at embryo adoption. We didn't want to pursue either option.

More than a year had passed since the last miscarriage, and by then we had decided to be childless and that my body had chosen not to be fertile any longer. Thus it was a shock when I found I was pregnant again in April 2004. There was implantation bleeding, strong morning sickness, and bleeding at 9 weeks (pink), but the ultrasounds matched the conception dates, and all sizes and growth looked healthy. A true miracle. I did take a low dosage of baby aspirin at the beginning of this pregnancy but stopped when I was bleeding. The doctor also placed me on Crinone. Otherwise I just took my prenatal vitamins and that is all (I quit exercising after the bleeding). In December 2004 I delivered a healthy boy after 12 hours of labor. He is completely healthy and happy.

Another miracle happened in our lives around the first trimester of that pregnancy (when I was sure I would miscarry again). We received a call that a birth mother had chosen us to adopt her newborn infant (we got her at 10 days) after being on a waiting list for 2 years. We finalized the adoption within 6 weeks!

I'm not saying it wasn't stressful being pregnant and having a newborn, but it is a miracle to have two children (7 months apart). We feel very blessed and know this completes our family, because I am over 40 years old and don't feel the need to push my luck farther. Our children bring so much joy to our lives, and I am so glad I persisted until I was successful in being a mother. They are our joy and bring so much fulfillment to our lives.

Lisa

I was in my first pregnancy and due in April 1997, just shy of my 30th birthday. My husband and I had been married for 3 years and thought perhaps we should start "trying" for a family, not knowing how quickly we'd be successful. I did a home pregnancy test 2 months later, and we were pregnant. We were excited and nervous and wanted to tell the world.

My first trip to my doctor confirmed the pregnancy with a urine test. We bought *What to Expect When You're Expecting* and started following the progress of our little baby. For Thanksgiving we brought a decorated cake to the in-laws' farm that read, "Happy Thanksgiving Grandma and Grandpa" as a tasty way of sharing our big news. This little one was the first grandchild for my husband's parents and the first in 22 years for my parents! We were all so excited we were silly with happiness.

Week by week we read about the development of our little fetus. I had been feeling pretty good but was definitely needing to nibble on crackers to keep my stomach settled. As my tummy grew, the nausea increased, and as the third-month mark approached, we began telling friends and coworkers, one by one, about our wonderful news. It was so much fun. Everyone was so happy for us.

Our small city had no maternity stores, so a trip to "the city" and some shopping with my big sister produced my first real maternity clothes and the first bits of baby stuff. That shopping trip was a difficult one because my stomach was so upset I had to look for the washrooms the minute we entered each store, just in case. Sometimes I was fine, but too often I was sick to my stomach. This "morning sickness" was throwing me for a loop! Otherwise I felt great, and I had read that this queasiness would eventually pass.

I had regular trips to my doctor where she checked my progress and nutrition. We had not yet been able to listen to the baby's heartbeat, but my doctor said she wasn't always able to track the little one down and not to worry. We'd do it at the next visit.

One day I noticed a tiny bit of blood on my panties. It was a quiet little problem, but of course I was concerned. I popped in to a walk-in clinic after work that day just to ease my mind. The doctor on call was a young woman who was upfront with me. She said that they think that as many as 30% of all pregnancies end in miscarriage and that mine may be one of them. She suggested I make an appointment with my regular doctor. I nodded, thanked her, and left. By the time I reached my car I was in tears. I drove home and told my husband that things might not be okay. We were worried and trying to be strong for each other, but we were also still quite optimistic; we certainly were not ready to believe a miscarriage was happening. I still had so much morning sickness that I felt very much pregnant.

A few days later I saw my doctor, who again checked for a heartbeat without success. She said again not to worry and that she'd schedule an ultrasound just to ease our minds. We left feeling somewhat better at the prospect of having an ultrasound. It was such a pregnancy thing—the ultrasound! This big event was at once both exciting and nerve wracking. As I continued to struggle more and more with morning sickness, we waited for the day of the ultrasound.

On October 9 we went to our hospital for the ultrasound. As every mother knows, I was to drink a ridiculous amount of water ahead of the appointment so that the pictures would be nice and clear. I was so nauseous now that I could hardly function. We no sooner arrived at the ultrasound department than I had to run to the washroom and throw up all the water. I was so worried they would tell me they couldn't do the test and send me home, but they took me in. I was laying on my back making small talk to the nice technician. He was asking questions about how far along I was, and so on, but it gave me the feeling something was wrong. He then said he needed to consult with his supervisor. My heart dropped a bit more, and I knew—but could not yet believe—it was bad news. He returned with his supervisor, who viewed the screen and agreed with his diagnosis. They then called my husband into the room and told us that they were unable to find a heartbeat or a fetus. They gave us kind looks as they said they suspected I had had what was known as a molar pregnancy, and they would notify my doctor immediately. I said we had a scheduled appointment the next day and they told us to keep it; she would be able to tell us more. They told me that I could take Gravol to settle my stomach. I hopped off the table, got dressed, and met my husband in the waiting room. We linked arms, and I tried really hard to hold my head up and the tears in. I know my face must have reflected how I felt. I knew we were walking past people we knew but all I could do was concentrate on getting out of there and making our way down that very long hallway to the daylight that indicated the exit. I think we were about 10 feet from the door when I started to sob. I sobbed for a very long time, and so did my husband.

The next day we went to the doctor's appointment together. The minute we stepped in her office she hugged me, and I broke down. I cried all the way through the appointment, but I will forever be thankful for that hug. That one act of human kindness that doctors often don't feel comfortable giving meant the world to my husband and me. Throughout the coming months, my doctor kept in touch (even calling us at home over the Christmas holidays) and made sure that both I and my husband were doing well physically and emotionally too. She's a gem.

My doctor had never treated a patient with a molar pregnancy before. She explained as much as she could and answered many of our questions. She said that she'd arrange a D&C for me immediately after the long weekend. She was wonderful, but we still had many questions. Being able to take medication for my nausea made me much more comfortable, but the knowledge that I had a whole lot of nothing in my tummy was a bit sickening. I remember thinking that it would be worse having to carry a dead baby over the weekend, that I should feel…better?…somehow knowing there never was a baby. I didn't. It was a very long, difficult weekend. As every parent who has experienced a miscarriage knows, you lose your hopes and dreams for a beautiful future with the loss of your baby. Your world is turned upside down, and your heart breaks.

On Sunday night I met my specialist at the emergency room, where she implanted a laminary tent in my cervix. This little stick would absorb moisture and expand, which would somewhat open my cervix in preparation for surgery the next day. It hurt when she put it in but not afterward. On Monday, October 15, my husband and I went to the hospital for the D&C. I remember hoping the nurses didn't think I was having an abortion. Not that I've ever been against abortion, but I wanted this baby. I wanted to be a mom. Somehow that mattered to me. My husband held my hand in the day-surgery prep area. I could see the hurt in his eyes. The surgery went fine, and I felt well afterward. I went home and had no discomfort or nausea. Physically I felt good, and emotionally I suppose I was over one hurdle and trying to look forward.

The next week I had my blood drawn to measure the level of HCG in my blood. This "pregnancy" hormone was what had been making me so ill. My levels at the time of my surgery were 560,000—a number that I later would learn was quite high. The biopsy report noted the tissue did not appear to be cancerous and that I had tolerated the procedure well. Each week I would have my blood tested until no more HCG was in my system. So, on Monday mornings I went to the lab and chatted to the patients in the waiting room and then to the nurses who added a new bruise to my arm. The first test showed that with the removal of the tissue the levels had already dropped to 14,000.

After only 2 weeks of blood tests my levels hit a plateau at 7,600. "Not uncommon," I was told. But the following week my specialist called me at work and quickly rattled off that "it" was back and that she had made appointments in Winnipeg for that Friday. I would most likely be starting a course of chemotherapy on Monday at the Cancer Research and Treatment Centre. Then she hung up. I am the dean's administrative assistant at a small university, and I spend my days recording information. I managed to write down most of what she said, but I had people in my office waiting to see the dean and me. I could feel the blood draining from my face and my knees getting weak. I sat down before I fell down, hung up the phone, and answered the questions of the person in my office. Then I grabbed my purse and left, mumbling to a coworker that I had to leave. All I could think about were the words *chemotherapy* and *cancer*. CANCER. I didn't know what to do or where to go. My husband was at work, and my parents were only blocks away from the university. I landed on their doorstep and burst into tears in a confused, emotional outpouring of information. I remember having to tell my elderly parents that I didn't know if I HAD cancer or what all of this meant. I didn't know if I could have kids. I didn't know how scared I should be. I didn't know nearly enough. I remember the shocked looks on their faces, and I remember the parental look of "we'll get through this." They listened, and I calmed down. I don't even now remember telling my husband that night and neither does he—we were in shock.

Those were pretty emotional, confusing days. No one I knew had ever heard of a molar pregnancy; none had experienced one themselves. My boss's wife was a nurse, and I asked her to see what she could find out for me. I searched the Internet and only came up with clinical references to treatments, medications, and so on. There was nothing in layperson's terms that answered my questions. The "not knowing" was horrible. I knew I'd been given a sentence but didn't know the extent of the bad news. The possibility of not being able to be a parent had never occurred to me before. I'd always been sure that being a parent would be part of my future. I had always known that. I couldn't imagine now what kind of a future I would have. What did this mean for me? For us? Would my future be short? I just didn't know.

Our first meeting on November 14 with the oncologists at the Health Sciences Centre in Winnipeg—the capital of the province and a 2½-hour drive from our home—was overwhelming but encouraging. I met doctors who gave me answers to almost everything I asked. The entire facility knew about people like me, people who needed chemotherapy and who were in a foreign and scary world. There were helpful and compassionate faces at every desk. People actually *walked* us to the correct offices instead of just telling us the way. They seemed to understand that we needed extra help.

I remember distinctly the doctor telling me to go back on the birth control pill because it might offer some protection for my ovaries from the chemo. No, I didn't need to think about freezing eggs. No, my chances of being a parent were still fine. However, the chances of me having another molar pregnancy were higher than for the average woman. After I'd completed my chemo, which would take approximately 3 months, we could start trying for a family again in 6 months. He said, "if you're going to get cancer, this is the cancer you want." It was curable. VERY curable. There was a very high success rate. And just before he left the room he said, "Oh, and that 'why me?' question?...Don't worry about it. You'll never find an answer and you'll just drive yourself crazy looking for one."

I hadn't even recognized that question within myself yet. There had been so much to deal with: the loss of a child and the loss of the happy future that we had been planning; a surgery; having to tell others what had happened and dealing with the looks of pity and sorrow from other people; dealing with our own feelings of sorrow; the stress of the unknown; the quick trip to Winnipeg to a medical facility we'd never been to; sick leave from work; finances.... I had also been carrying with me this sense of betrayal. I had prayed to God for a healthy baby (I never asked God for anything), and instead he gave me a tumor, possibly cancer. What a slap in the face from my God! What had I done to deserve being faced with death when I had hoped for a new life? When the doctor mentioned the "why me?" question, I broke down. It HAD been in the back of my mind. I just hadn't the time or courage to ask that question yet. While it would take many years to really get over things, at that moment a lot of air went out of the balloon. It was no longer something pressing from back in my subconscious. His words helped me tremendously, and I never thanked him for that.

We went home to Brandon and planned the next week in Winnipeg. My big sister lived in Winnipeg, so we had a place to stay. That Monday, my Mom and I headed to Winnipeg because my husband's company wouldn't give him the time off work. It was just as hard for him to watch me leave as it was for me to go. There is a bitterness toward his employer that both he and I still carry. We have not forgotten that unkindness, and from that day forward his dedication to his employer diminished. He still goes to work for the paycheck, but nothing more.

My treatment was chemo (methotrexate, actinomycin-d and lasics) every day in Winnipeg for 5 days, Monday to Friday, and then home to Brandon for a week of rest. Then back to Winnipeg for a week of treatments. Back to Brandon for a week of rest. The nurses were angels in disguise (for years I sent them Christmas cards), not so much for me, but certainly for the other people receiving chemo. I was in good shape. Most were not. All the breast cancer patients were bald and vomiting. Many of them had a private room and some would even stay overnight. The rest of us sat in recliners in an open room with our IVs in our arms. When I first arrived for the week they checked my beta HCG levels and other indicators of my health/tolerance through blood tests. Then I was off to the treatment room, where I soaked my arms in warm water to help them find a good vein for my IV. The treatment itself lasted about 2 hours. They would then put heparin in my IV connection, which allowed me to go home with it that night and saved me having to get another IV in the morning. This is called a heplock. Although it was not the most comfortable thing, and I did learn to hate it (and all needles) toward the end of my treatment, it wasn't the worst thing either. I wrapped my arm up in one of my 6′5″- nephew's tube socks at night and showered with a bag over it, and that was that.

The other patients at the center ranged in age from small children to senior citizens. When the children came in, they really livened up the place. Regardless of their medical situation, kids are kids. They yelled, laughed, and behaved like kids. They were wonderful for my spirits. I don't think I could have that attitude now. I'm afraid I would see an innocent child's suffering and not be able get past that. I'm afraid it would just break my heart. Some of the other patients shocked me. Those with lung cancer would sneak outside for cigarettes. Or worse, if they were too sick to make it outside, they would smoke in the washroom of the Cancer Treatment and Research Centre! What a powerful addiction.

The volunteers made each day so much better. On most of my treatment days, I would be picked up at about 9 A.M. by a volunteer driver. Almost all of my drivers had been cancer patients at one time or another. Some were family or friends of past patients. All were amazing. Both my sister and her husband worked and were not able to drive me to my treatments or pick me up. Without these wonderful volunteers I would have had to handle the drive and parking by myself or run up taxi fares. You could also have honest and frank conversations with them about your situation, and they understood. They weren't shocked, and they never said anything tinged with pity. They also drove me home at the end of my treatment. It was a wonderful to have one less thing to worry about each day.

There were also volunteers who came around with juice and cookies and smiles—a nice thing to look forward to each day. One memorable day a small group of players from the Winnipeg Symphony Orchestra serenaded us all morning. It was a few weeks before Christmas, and it was lovely. The room was filled that day with a tranquility and lightness. I have no doubt it helped us heal. It was beautiful.

Then there was my sister and her husband. They opened their home to me, pampered me when they could, and took great care of me. They made the ordeal SO much more comfortable and filled my days with much needed laughter. I could not imagine having to be there all alone, getting my own meals and driving to and from appointments. This support group made every difference to the experience.

On my weeks home in Brandon I worked a few days. I was usually feeling crappy at the beginning of the week but fairly well by Wednesday. I did, however, use up all of my sick time in the 3 months it took to complete my treatment. During those months I couldn't find the courage to tell everyone about my ordeal. I could and did tell them that I was taking chemotherapy, but I couldn't get out exactly why. I just could not bear to see the looks on their faces when I told them that I had a lost a baby and now had a cancer-like condition because of it. I couldn't get a good grip on those details myself. I cried each time I spoke of it, and I was too proud to cry in front of most people. Unfortunately, rumors started about what exactly was wrong with me, and there were a lot of misconceptions among my coworkers and more distant friends. Some people thought I might be dying, and others were outraged that I wouldn't be open about my illness. I suppose they thought I was hiding something (and I was).

My treatment progressed well, I suppose. My beta HCG levels were 11,867 on November 22 when I started treatment, and they continued to drop week after week. The antinausea medications prevented me from being sick. Most days I just felt like I had the flu. I do, however, remember that the levels decreased at such a painfully slow pace toward the end. On December 5 I had a count of 348 (better!), and on December 16 it was down to 73. I spent that week in chemo writing up our Christmas cards to family and friends. I remember the sincerity that I felt every time I wrote "wishing you good health and much happiness." Really, what more could anyone ask for? During the 3 months of my treatment I was filled with an overwhelming sense of gratitude for all I had in my life. I was blessed. I had friends who cared about me, a wonderful husband, and a family who loved me deeply. I had no interest in the small inconveniences of life. I had time to slow down, think, and appreciate. I was blessed.

By December 19 my levels were at 40; December 30, 7; but on January 2 I was still at 6.2. Auuggh! It took one more week of treatment before I would finally be done chemo. The doctors said we should have a 6-month waiting period to ensure that the HCG levels remained where they should be, and then we could start trying again for another baby. My due date and my 30th birthday came and went (not the happiest days for me), but by spring I was feeling great both emotionally and physically.

In August, the month after we started trying, I became pregnant. Again it happened so easily that I was both thankful and scared that we would be headed down the same path. My doctor scheduled as early an ultrasound as made sense, just so we all would know the situation. Close friends knew we were expecting, but no others. I cannot tell you how nervous I was at that ultrasound appointment. I put on a brave face and chatted to the technician. They remained fairly quiet, pushing buttons, moving the sensor around my belly. I tried to casually ask if everything looked normal. They replied yes, that they were just taking measurements, and when they realized why I asked, they apologized for the delay in letting me know. I cried and cried with joy, and my belly bounced so much that they had to stop. They just smiled and let me cry.

My pregnancy with "Cletus," as my niece nicknamed our baby (Curtis and Lisa's fetus) went fine. His delivery did as well; he was born May 23 at 1:59 P.M. after only 5 hours of labor. He was 8 lb, 9.5 oz, and 21¾ inches long. He was beautiful. Perfect. Our miracle. Two and a half years after "Cletus's"—Tyler's (his given name)—birth, we were blessed with our beautiful angel MacKenzie Grace. Kenzie was born in 3 hours (ouch!) and was 6 lb, 13 oz. Right from the moment she was born she has been gorgeous inside and out.

It has been 10 years since that molar pregnancy, and for me, all that remains is a sense of unease or foreboding that washes over me each fall when the leaves change and there is a chill in the air. It has been like that every year since the diagnosis. Our children are healthy and happy and full of love. Tyler is 8 and Kenzie is about to turn 6. We are blessed.

I had always felt, during the bumps in life, that things happened for a reason. And I had always been able to find the reason relatively quickly. I'd always felt that there were lessons to be learned, something better ahead, a need to change my direction. But for years I've never found a place for this experience. I learned that I am very strong, but I didn't see that as something I didn't already know about myself. It wasn't until one day this last summer, when my husband and I were sitting back watching our children play, that perhaps I found my answer. We had both agreed long ago that we wanted two children. He commented on what a joy our daughter Kenzie is and said, "You know if that first pregnancy had worked out, we never would have had Kenzie," and I cried one last time.

Lisanne

On December 25, 2002, I conceived my first child at the age of 39 after trying for about 15 months. I did two home pregnancy tests on January 13, 2003, and they were positive immediately. My husband and I were elated. We made the decision to not tell our families until after my first visit to the doctor. My doctor had told me he always does an ultrasound during the first visit at 7 weeks. My appointment was on February 5, 2003. We planned to tell our families by showing them the ultrasound pictures the weekend of February 7. The doctors also thought I might have a good chance of having twins, because my mother is a twin and I am older and this was the first pregnancy. I was thrilled at the thought of having twins. It was what I had always wanted.

Finally February 5 arrived, and my husband was in the room as the doctor began the ultrasound. I had a sick feeling in the pit of my stomach that something was wrong. The doctor asked me again when I had had my last period, and I answered December 13, 2002. He then told me that he could see no sac with a baby and there was no heartbeat. However, he did see a large cyst on my right ovary. I was completely devastated. I felt as though my world had fallen apart. I completely lost all control right there. He took us to his office and told us he was sending me to the lab to get my HCG levels checked; he thought maybe I wasn't as far along as I thought. He promised he would let me know within 3 hours. I was unable to return to work because I couldn't control my emotions. I got on the phone and called my mother. It was quite a shock, since she didn't even know I was pregnant. I was such an emotional mess that she told me that she and my sister would travel the distance to be with me.

In less than 3 hours, my doctor called me and told me my HCG levels were at 40,800, which meant I could be up to 12 weeks pregnant. He was trying to get a specialist to do another ultrasound that day if I could go, and I told him I would definitely go if he could get the appointment. Within 10 minutes, he called to say he had located a doctor who would do it within the next couple of hours. My sister and I went to the appointment, and the first technician began by doing a regular ultrasound. She could not see a sac or a baby. She then did a vaginal ultrasound. She still didn't see a baby or a sac. Another technician came in to view the ultrasound. She didn't see anything either. They then called in the doctor, who had been outside the room viewing the pictures as the ultrasound was being done. He came in and looked at the live ultrasound and told me that he was almost sure that I had had a molar pregnancy. He said most of the time the mass will look like a cluster of grapes in the uterus, but that mine were quite a bit larger. He called my obstetrician immediately and told him what he thought. My obstetrician asked to speak with me and he told me that sometimes a molar pregnancy can turn into cancer, but that if I would agree, we would wait another week just to be 100% sure it was a mole. He promised that a week would not put me at more risk. I kept thinking maybe they had made a mistake. I chose to wait another week in the hope that some miracle would happen and all this would be just some bad dream.

It seemed as though the week would never pass. It was total misery for me. I couldn't concentrate on anything, and all I did was cry. I prayed and prayed that everything would be okay. I wanted this baby so badly. Everyone tried to comfort me by saying that everything happens for a reason and that it was better now than later. They meant well, but they weren't losing a child. On Friday of that week, I got the nerve to go on the Internet to research molar pregnancy. I thought I should educate myself. It was complete horror. Everyone said I shouldn't be reading this stuff, because most of it probably wasn't accurate. I told them I had to know something. Finally, the week passed and my husband and I went back to my doctor on February 12, 2003. He had his assistant do another vaginal ultrasound, and she didn't see a baby or a heartbeat. He came in and tried again himself, but still nothing. It had gotten larger during the week, and he was now pretty certain that it was a molar pregnancy. He told me that I should have a D&C and that I would need to have blood drawn on a weekly basis for 6 months. He also told us that the timeframe to get pregnant again would be 6 months instead of the 1-year waiting period suggested by most of the articles I had read. He told us the risks involved in having the D&C and the risks associated with the molar pregnancy. He said that it appeared to be a complete instead of a partial mole but that the pathologist would make this determination after it had been removed. Because I live in the Raleigh area, he told me if I wanted he would send me to Duke University because there was a doctor there who specializes in molar pregnancy. I chose not to do this because I trusted this doctor and did not want the waiting time to be prolonged. I knew from just waiting 1 week that I couldn't handle waiting any longer. I asked him if there was any way he could do the surgery the following day. He called the hospital and made the arrangements for noon the next day. At this point, I knew there was nothing the doctors or I could do. Although I felt hopeless and helpless, this was my only choice. I decided that I had to put on my tough face because I knew the journey ahead would be a long and rough one.

On February 13, 2003, I made my way to Rex Hospital in Raleigh, North Carolina. Everyone was very kind and understanding of my situation. The doctor ordered lots of blood tests and chest x-rays before the surgery. He wanted to go ahead and check my liver function and check my lungs to see if it had spread. He also checked my HCG levels again. Before they put me to sleep, they put the blood bank on alert for a transfusion in case I lost too much blood. He also told me that my lungs looked okay but that my HCG levels were high—now 120,000. They had tripled in a week. Fortunately, the surgery went well, and I didn't have much bleeding at all. I left the hospital 3 hours after surgery. I felt tired and still had all the symptoms of the pregnancy, but overall I felt pretty good physically. Emotionally I felt as though I were dying inside. I tried to keep my tough face on for my family and friends. I knew that I must grieve in my own way. Unless someone has been in my shoes, I'm not sure they understand my feelings.

The day after the surgery, my doctor called to say the pathologist had diagnosed it as a partial mole, instead of a complete mole, which was better for me because the cure rate was higher in partial molar pregnancies. I had my first HCG test 2 days after surgery, and my levels were down to 20,000. After doing well for a few days, I started having pain in my left lower back and side and pain in my chest. I even had some shortness of breath. The pain got so intense that I began to vomit. I called my doctor's office, and one of his partners told me to meet her at the hospital. They again did blood tests and chest x-rays. The chest x-rays showed quite a bit of inflammation in my lungs, and the doctor said they sounded like a foghorn. They hoped it was just pleurisy or a touch of pneumonia, and they thought the pain in my side was coming from a fibroid tumor they had found on my uterus. My HCG levels were dropping rapidly, which is what they had hoped for; by February 19, 2003, they were down to 4,000.

I went for my postoperative visit on February 21, 2003. My lungs were clearing up with the antibiotics, and my uterus was down to regular size. I was actually beginning to feel a little better physically, but the emotional pain wasn't healing as quickly. Two weeks after the surgery, my HCG levels were at 581 and still falling. On March 5, 2003, I went for my weekly blood draw. I had been feeling really sick during the week. I received a call from my doctor the following day while having lunch with my husband. I could tell by the sound of his voice that the news wasn't going to be good. Unfortunately, my levels had increased to 1,699, up from 581. I told him I already knew. "How do you know?" he asked. I told him I was getting sicker every day. He told me he would have his office call me later that day with an appointment to see the leading specialist in this field at Duke University. This was on a Thursday. On Saturday, I had to have a CT scan of the body and chest x-rays to take with me to Duke on the following Monday. I had no time to fall apart. It was like a whirlwind trying to get everything in order. I knew from my research and from what my doctor had told me that chemotherapy and cancer would be a part of my existence. I went home to visit with my family and friends after the tests on Saturday. On Sunday, my mom came home to be with my husband and me.

On Monday, my husband, my mother, and I made our journey to Duke. We saw the doctor and met with a chemo nurse. I was then sent for another ultrasound. My HCG levels were tested again, and before I left Duke we already had the results. My levels were now at 2,385. They talked to us about what to expect with the chemo and asked me to participate in a study on gestational trophoblastic disease. There was so much going through my mind, I asked if I could give them an answer about the study the next day. Being in a blind study such as this leaves you not knowing which chemo drug you will get until you arrive for the first treatment. There are only two different drugs involved in the study. The following day I called to let them know I would participate with the hope that one day it may help someone else with their struggles. It would be only 2 more days before the actual chemo would start.

I went to work to try to keep my mind off what was actually going on. It was a struggle. By the time I arrived at Duke on Thursday, I was quite scared. I went in to select my drug, and it was going to be methotrexate. My nurse told me she was glad that that was the drug selected because it was more widely used. She took me to the treatment room to introduce me to the staff who would be in charge of the actual treatment. I remember looking around and seeing all these really sick people and thinking, "Is this what I'm going to be like and look like when this is over?" In treatment rooms, everyone is just lying around in recliners and beds. There is no privacy. Because my chemo was an injection in the hip muscle, I was taken into a bathroom for a little privacy; all the beds with curtains were already full. It was so horrifying. The injection hurt really badly and began to bleed, even onto the floor. I started to pass out, so they took me to a recliner to rest. I remember seeing my husband when I walked out, and he had a look of shock on his face because I was so pale and couldn't walk without help. After sitting a while, I had to be wheeled out in a wheelchair to the car. I couldn't believe my life had come to this.

In the weeks that followed, I was always given a bed in the treatment room. They began to give me IVs, steroids for energy, and more antinausea medicine in addition to the chemo injection. They tried everything they could to make me more comfortable and less sick. After weeks of all the medication, with really no success, I asked them to just stop the medications and do the chemo injection. They agreed, because the medications were really not helping me. When people asked me how I was doing, I would just say, "I'm hanging in there." It was all I could do.

For the first 7 weeks of chemo, I was sick pretty much every day. I was totally miserable and had many side effects. I also continued to have morning sickness, just as though I were still pregnant. The first thing I did every morning was vomit. Although I was taking eight different antinausea medications, the sickness continued. The doctors thought maybe it wasn't just the chemo making me sick, but the HCG levels. They said my body just would not accept that I was not pregnant. My levels continued to drop, but at a slow pace. When I began my chemo on March 13, my levels were at 3,130. At week 2, they were at 965; week 3—656; week 4—166; week 5—117; week 6—114; week 7—49; week 8—25. Some weeks we saw good drops and some weeks they were not so good. Between weeks 2 and 3, I became totally devastated. I thought I couldn't handle it anymore. All I could do was cry. I was a total wreck. The doctors, however, tried to convince me that any drop was good. By the eighth week I started to feel like some progress was being made. My levels had finally dropped enough that my body accepted that I was not pregnant, and the sickness stopped. I was only taking a couple of medications at bedtime. The doctor also said my body had become accustomed to the chemo. The doctor and nurse told me that even when my levels finally dropped to below 5, I would still have to endure at least two more chemo treatments. After that, I would continue to have my levels checked on a regular basis, beginning with once a week.

I was lucky during this time that I had a great support system, including family, friends, and coworkers. God blessed me with the best husband in the world, and my mother is the most wonderful mother a person could ask for. There was also another very special person in my life, a coworker who has battled breast cancer recently and had to have chemo and radiation. She was a real friend to me on good days and bad days. I could always tell her anything. We talked about our battles and how we were feeling. She was truly someone who could relate, even though our diseases were very different. The medical profession doesn't prepare you for the emotional toll of chemotherapy. I will always have a real bond with her, a bond that will never be broken. I will never be able to thank her enough for standing by me.

During week 9 my levels dropped to 16, but at week 10, they hit a plateau. They were still 16. I began to get very discouraged again. I even thought of stopping the chemo altogether. It was just getting to be more than I could bear. I told myself I would give it one more week. I knew if it didn't change, I would be administered another chemo drug, which would have even more side effects. Week 11 brought my levels down to 12. It wasn't big in numbers, but in percentages, it was 25%. I was grateful it was working again. I now felt I had hope again. During week 11 I also began to have shortness of breath and a nagging cough. It was quite scary, thinking the mole may have spread to my lungs. X-rays were ordered, and the doctor examined me and found that I had a slight respiratory infection. The cancer had not spread. I was very relieved. Over the next week, I started having a lot of cramping and bleeding so I called the nurse. She said this could be a good sign. It indicated the molar tissue was finally releasing from my body. That was exciting news to me. When I went for my 12th treatment, I was examined by the doctor, who confirmed that it was molar tissue. Before receiving my treatment that day, my husband and I got the news we had been waiting to hear for all these months: My levels had finally gone down to 5, which was the magic number! Five or less was considered to be negative. I received my treatment and went home knowing that I only had one treatment left. As week 13 approached, I was very excited and anxious for chemo day to arrive. On June 5, 2003, I had what I thought would be my last chemo treatment. I was thrilled to have this part of my terrifying journey over. I went to chemo and upon leaving gave my favorite nurse a hug and told her I would not miss going to chemo, but I would miss her great personality. I hoped I would never see the treatment room again, unless to visit her.

However, the bad news came shortly thereafter. My levels went up again. Devastation was not the word for how I was feeling. I was told just to wait until the next week and have my levels checked again, and we would have to make decisions from that point. On June 12, 2003, I got my results, and my levels had dropped to negative again. When my nurse told me, I had a brief moment of excitement. Then, once again, I got the bad news: I had to start chemo again. I didn't even get to skip a week. Because the levels were elevated one week and negative the next, I had to have at least two more treatments. I was dreading it. Even after the chemo finally ended, I would still have to spend the next year of my life having my blood levels checked and waiting around for the results to see if the cancer comes back.

On June 20, 2003, my levels were negative for the second week in a row, which meant I could stop the chemo. After 15 treatments, I was physically and mentally exhausted and now had to start to rebuild my life. By July 25, I had had five blood tests that were less than 5, and I was allowed to switch to biweekly blood tests. I was also beginning to feel a little better physically. Chemo kills the good cells as well as the cancer cells, and after chemo, it takes at least a few months for the cells to rebuild themselves. I had hoped when chemo ended that I could get away for a few days at the beach, but I chemo damages skin cells as well, and the doctors said I would probably burn and would risk having melanoma. It was not worth the risk! I hope I never have to see another chemotherapy treatment room. Since finishing chemo, I had not been able to make myself return to Duke. The memories were still too fresh in my mind. I had my blood draws and doctor visits at a satellite office in Raleigh.

By September 26, 2003, my levels had stayed below 5 since June, and I was finally having my blood drawn only once a month. I had also finally regained most of my energy. Our baby would have been due on September 18, so that week was difficult for us emotionally, but we got through it, just as we have gotten through everything else. I still had 9 months of blood draws left before we could even *talk* about having another baby.

In June of 2004, I reached a milestone. It had been 1 year since my last chemotherapy treatment, and we were given the okay to try for another child. My husband and I immediately sought the help of an infertility specialist. We did lots of tests and even did a laparoscopy to remove some scar tissue. The doctor told us we would never get pregnant on our own because I had a blocked tube, probably caused from a ruptured appendix that occurred in 1989. However, he agreed to try me on Clomid for 1 month. I did produce eggs, and I became pregnant in September 2004. Almost immediately, I began to miscarry. Once again, my husband and I were devastated. We thought we could not endure yet another loss. After this miscarriage, the doctor continued to insist we would never get pregnant on our own. He kept pushing us to try in vitro fertilization (IVF), which we did not want to do. We had informed him of this during every visit. He had already told us there was only a 3% chance of getting pregnant with the IVF and that we would have to travel out of state to do the treatment. After a few visits, it became apparent that he was in it for the money. He knew we were desperate, and he played on our emotions, but we were able to see what he was doing. When I returned for my postoperative visit I informed him that we no longer desired his services and that I had been doing some research on my own and had made an appointment to see an infertility specialist at Duke. She had come highly recommended and specialized in egg donor programs.

I went for my visit with the specialist, and although she looked at the same test results as the previous doctor, she did not agree with him at all. She said I could become pregnant and that we should try again on our own. We decided to give it until March 2005 before pursuing the egg donor program completely.

As luck would have it, I conceived again on January 19, 2005. In February 2005, I had a positive pregnancy test. We were happy beyond belief. We thought we were getting another chance at our dream of a child. I immediately called the doctor who had been my obstetrician during my molar pregnancy, whom I still believe is the greatest and most compassionate doctor ever. His staff immediately scheduled an appointment for me. On the ultrasound we could not see a heartbeat, but we could see a normal-looking sac. The doctor told us that it still could be a little early, because I was only about 5 weeks along. He advised us to wait 2 weeks and then return for another ultrasound. Those were two of the longest weeks of our lives. On the second ultrasound, the heartbeat immediately was apparent. I let out a loud yelp. My husband and I were elated. The doctor printed us out lots of ultrasound pictures, and we shared them with all of our family and friends. I was 7 weeks pregnant and our baby was alive!

This was on Friday, March 4, 2005. On the following day, I was cramping a little, and by the next day I had a little spotting, but I knew this was sometimes normal in early pregnancy. By Monday, the pain had increased. I called the doctor's office, and they told me to go on complete bed rest, which I did. On Tuesday I felt even worse so I called back, and they told me to come on in for another ultrasound. My usual doctor was not available on this day, so I saw one of his associates. As soon as she started the ultrasound, we saw there was no heartbeat. She confirmed this and also called another doctor in to look at the ultrasound. She told us our baby was now dead. The doctors left the room to give us time to compose ourselves. When she came back in, she told me that I needed a D&C and that they could do it the following morning, so surgery was scheduled for March 9, 2005. She also arranged to do genetic testing on the fetus in the hope that we could find out why we had just lost another child. We went home, and an hour later, as I stood up, blood gushed out of me. I immediately called the doctor's office, and they told me to get to the emergency room. The doctor examined me and said that I had expelled all of the tissue and that I needed an emergency D&C, so I was taken into surgery. A few days later I received a very disturbing call from the doctor. She said she was afraid that all the tissue may not have been removed because there was not enough tissue for the genetic testing she had ordered. She told me I must start having weekly HCG tests done to make sure my levels dropped below 5 and to make sure I didn't have yet another molar pregnancy. We were now struggling with the loss of our third child and the possibility of my going through living hell again. Fortunately, within a few weeks, my levels were normal. We were so relieved. We would start over once again.

By May of 2005, I was pregnant again. I immediately went to my doctor and had an ultrasound. I was only about 5 weeks so we saw no heartbeat, but the doctor decided to start doing my HCG levels every couple of days to make sure it was doubling, rather than wait for another ultrasound. My first HCG was done on May 15, 2005, and it was 371. When repeated on May 17, 2005, it had already started dropping. My doctor called this one a missed miscarriage and told me that instead of putting me through another surgery, he suggested that I just let nature take its course. By Saturday, May 20, 2005, I began to cramp and bleed pretty bad, and yet another child was gone. This was our fourth little angel.

I continued to have my HCG levels checked until they were below 5. After my levels dropped, my doctor called me and asked what my husband and I wanted to do about trying again. I told him we would really have to think long and hard, because we had suffered way too much heartbreak. He told me he saw no medical reason why we shouldn't try again. He also informed me that none of these losses were a result of the molar pregnancy. He told me if I did become pregnant again, to immediately start taking one baby aspirin per day, and he would also give me progesterone shots.

Over the next few months, we didn't really try to become pregnant but we also didn't try to prevent it. In July of 2006, we sold our house in the Raleigh area to relocate near Wilmington. Somehow, during the relocation and the starting of a new job, I lost track of my last menstrual cycle. By mid-August I started to feel sick, but I just passed if off as the result of stress from all the changes in my life. However, on August 21, 2006, I did a pregnancy test, and it was positive immediately. I was happy but also very scared. I began to try to find a doctor. I called four practices, and when I explained my situation, they all refused to take me on as a patient. I was horrified. I called another practice and finally got someone to hear me out. She obviously could hear my desperation, and she said she would see what she could do. I then called my doctor's office in Raleigh. I talked to the nurse and told her my dilemma. My favorite doctor called me twice and told me we would figure something out.

In the meantime, the other practice called me and said they would take me as their patient. I immediately went in, and the physician's assistant did an ultrasound. She saw two sacs and called the doctor in, but he only saw one, with no heartbeat. He told me it could still be too early and to come back in 1 week, but that we should remain very guarded. I asked him about checking the HCG levels for viability, but he didn't agree that it was necessary. The sac measured about 5 weeks on this visit. I returned in a week and still saw no heartbeat, but now the sac measured at 7 weeks, 3 days. He gave me three options: I could wait another week, I could schedule a D&E, or I could just wait and see if I lost it on my own. I chose to continue the pregnancy because I still had hope. I just couldn't give up because I had had neither pain nor bleeding. I still felt pregnant. I waited yet another week.

It was now September 5, 2006, and my husband and I went for another ultrasound. There was still no heartbeat, and now the sac measured only 6 weeks 4 days. The doctor spoke to us at length and told us that there was absolutely no hope left and that we should schedule a D&E and have genetic testing done on the tissue removed. Because of such a backlog at the hospital and surgical center, I had to wait 3 days. I prayed a lot that I wouldn't lose it at home and risk losing the tissue, because we really wanted some answers as to why this loss occurred. After five pregnancies and still no live birth, we had to have answers.

After about 3 weeks, the results came in, and it was more than one problem. It was a female fetus with only one sex chromosome, an extra #7 chromosome, and an extra #20 chromosome. Any one of these is considered to be fatal. The doctor informed us that he still could see no reason why we shouldn't try again. At this point, we have left it in the hands of God. We are exploring the possibility of adoption, because we have had no luck having our own child. We can only pray that God will see fit to give us a child. We will never give up that hope.

As I was finishing up this story, I got the results of my 6 month HCG test, and it was normal. So I can breathe for another 6 months. Next year I will reach the magic 5-year mark, and these will discontinue. It has been exactly 4 years today since I was diagnosed with cancer. To be honest, I was very hesitant to write again because I did not want to discourage anyone by my plight, but after much soul searching, I decided to go ahead. I wish all of you luck.

Marilyn

It was the strangest sensation in the world. I was running with a dead baby in me. Well, if there was a baby at all, it was dead. I had been to the doctor's office earlier for my 3-month checkup, and the doctor couldn't find a heartbeat, try as she might. The ultrasound showed only a "snowstorm effect," yet the doctor felt it was okay to tell me, "don't worry. It's probably just my machine; it's my job to worry for you." Your machine has suddenly become a meteorological instrument, and you're trying to tell me not to worry?!

Feeling numb, panicked, and on the verge of tears, I had to go sit in the waiting room with all the pregnant women for what seemed like forever while the nurse scheduled an ultrasound for me at the hospital. It wasn't until I got to the hospital that I was told it was to be an internal ultrasound. So, my baby was presumably dead, and I had to go have a horrible internal probe on a cold hospital table a few weeks before Christmas.

I called my husband when I got home from the doctor's office. Despite having been told not to worry, I sobbed into the phone. I had to pull myself together, though, because my 2-year-old needed me. When my husband got home, looking uncertain and not knowing what to say or do, I went for a run. Running was my way of coping. I went my usual route, thinking the whole while how horrible it was to be running with a dead baby inside of me. But the doctor hadn't exactly told me it was dead, she had left room for hope.

All hope faded, though, on that cold, cold, hospital table. I knew the news was devastating, based on the technician's silent clicking away at the buttons on her machine. When she was through, she told me to dress and then left for a long period of time to bring the results to whomever it is that reads them. When she came back in, she asked if my doctor was thinking along the lines of a molar pregnancy.

I had never heard that term before and, now crying, said I had no idea what my doctor was thinking. It was made all the worse by the fact that the technician didn't even try to reassure me or calm me down. My doctor happened to be at the hospital that morning and was coming down to see me. When she came in, she hugged me and explained that I had what appeared to be a molar pregnancy. I only remember bits of what she said. "The fetus died early on…the cells continue to grow…grapes or snowstorm…a type of pregnancy cancer can develop in some cases…call the office later today to schedule a D&C for next week." Wait a minute—next week? I couldn't walk around for the better part of a week with a dead baby inside of me! Perhaps it was the sobbing that did it, but the doctor managed to schedule the D&C for later that afternoon. She was very nice.

I called my husband from the ice cold car in the hospital parking lot. He hadn't come with me because someone had to stay with our son. I sobbed for a minute, then asked him to call my sister to see if she could watch our son while we came back for the D&C. I then told him to call his parents and let them know.

With my first pregnancy, I was 5 months pregnant before I told anyone. I couldn't hide it any longer and had to tell. I just didn't want to be fussed over and didn't want the attention. My friends had all had babies, and they had made such a big deal out of it and themselves that I just didn't want to be bothered. The second time around, though, I was excited to be having a sibling for my son. We had told everyone at 12 weeks. One day later, I was sitting in the doctor's office waiting to have the internal ultrasound scheduled.

We had about an hour between the time I got home from the ultrasound to the time we had to leave to drop off my son at my sister's house. It had snowed, and I decided we should all take a walk. I remember my son walked in his brand new shoes. I didn't cry. We dropped him off at my sister's house, and she said not to worry. She looked stricken. My son cried because he didn't want to be left, but I had to leave him. I had to go have his deceased sibling extracted.

Then the horrible ordeal began at the hospital. I had to wait and wait and wait for the procedure. Then my doctor came in and talked about having to order blood for the surgery. She also said that if it was required to save my life, she might have to do a hysterectomy to stop the blood flow. Suddenly, what I thought was a common procedure had turned into a potentially deadly one. I flipped out, and a very sweet anesthesiologist came and gave me something that would make me feel like I'd had a few glasses of wine. Despite the situation, whatever he gave me made me feel really good. The fact that I remember how good it felt despite everything that happened makes me understand how someone could come to abuse drugs.

I had never been put under before. I was terrified. I was also acting slightly drunk thanks to the drugs. I was thinking about the scenes from movies when someone was about to go under. They all tended to be very melodramatic. "I love you. If I don't make it...." I was thinking all that but didn't have the heart to say it. I was so worried about my son if anything were to happen.

I woke in a panic. My legs were shaking uncontrollably, I felt like I couldn't breathe and needed my inhaler, and I had the worst thirst I have ever known, but I wasn't allowed any water. Then I saw the two sacks of blood hanging from my IV and really flipped out. "You've almost killed me," I remember repeating more than a few times in the general direction of the doctors. Then horrible cramping kicked in, and I got a shot of something to ease the cramps. After what seemed an eternity, during which my husband and nurse actually laughed at my desperation, I was finally allowed a few ice chips. Then I got to spend the night in the maternity ward with all the new mothers and their babies. I was given a delivery room, off to the side, so I didn't have to listen to crying babies. It was all very surreal. I had lost a lot of blood during the D&C. What was supposed to be a 10-minute procedure had taken close to 2 hours. I still don't understand why. Later, after I switched doctors to someone that actually specialized in molar pregnancies, I asked him why I lost so much blood and if he thought it was possibly due to a botched job by the doctor. He responded that in cases like that doctors always "blame the patient." I gathered from his "sometimes patients have complications that are not normal" type answer that perhaps it was a botched job after all. I'll never know. But the implications of it are terrifying. Should I have another molar pregnancy, it is this potential blood loss that most horrifies me. Should I even have another normal pregnancy, the possibility of another "massive bleed" (as the pathology report described it) sends me almost into a panic.

I slept very little in the hospital that night. I remember the Pixies were on some late night show. I always loved the Pixies; now I have a hard time listening to them. I had an IV attached until late, but the line was kept open "in case." In case I needed more blood I supposed, but didn't want to ask in case of what. The IV hurt. The little blinking computer monitor made sleep impossible. I was so weak I had trouble getting up and walking a few feet to the bathroom. The weakness terrified me. I was a runner. I shouldn't have felt like I might collapse just walking a few feet across a room.

I had run during my first pregnancy. At about 4 months, my back started hurting to the point where I had to run a few blocks, then walk a few blocks. By month 5, it had become so uncomfortable that I miserably had to give it up. With this second pregnancy, the molar pregnancy, running was tough right from the start. It wasn't a discomfort, it was more a dead tiredness. I had trouble running 3 miles. My entire body felt like lead. It was not normal. I later learned that such extreme fatigue is characteristic of molar pregnancies. Extreme nausea is another symptom. I felt sick on a few occasions, but was more struck by my complete lack of an appetite. I never had any spotting, which is also another symptom.

The next morning, I cried a lot. I cried when the sweetest nurse in the world came in to see how I was doing. I was crying when the doctor on call came in. He was great. He explained what a molar pregnancy was. He explained how I'd need to get my HCG levels followed and would need to go on birth control. I don't remember what my questions were, but he answered them. He told me he usually had one molar pregnancy case a year and that in another 6 months, I'd be able to try getting pregnant again. At the time, it seemed like forever. Then he told me I shouldn't do any cardio work for at least 4 weeks. Again, the whole time was a bit of a blur, but I was comforted by his knowledge, a feeling I never got from my own doctor. I was also struck by the fact that he was a bit of a hottie—I needed something to help cheer me up!

While at the hospital, I received only two phone calls. Well, my husband called a few times, but he was obligated because he had to be home with my son instead of with me, so I won't count him. My friend called after the procedure to see how I was, but she failed to tell me that she had gone into labor. The next morning, after wondering if everyone had forgotten that I was hospitalized after almost bleeding to death (well, okay, I don't know if I almost bled to death, but a "massive bleed" in which I lost 1.5 liters of blood—almost a 2-liter pop bottle full! Yikes!), I finally got a call from one of my sisters. "So you had to show me up," she said. She had had a miscarriage at 5 months. I was only 3 months along, but the transfusion, night in the hospital, and cancer risk had her beat.

Another dear friend trekked all the way from the far north side of Chicago to the far south side to see me when I arrived home from the hospital. She just showed up to see how she could help. She was one of the maybe two people to ever inquire how my levels were doing as time passed. My neighbor also was absolutely wonderful. He walked our dog when I was in the hospital and when I was home but too weak to do it. He brought me a homemade pizza laden with iron-rich spinach and meat. He also brought over a bunch of beautiful irises. Then my in-laws arrived and put the irises in a small vase off to the side so that the vase above the fireplace could house the carnations they brought. They stayed at the house and just seemed to be in the way. It was very awkward and uncomfortable. I sent them outside to help my husband put up Christmas lights. It was still Christmas, after all, and my 2-year-old didn't deserve a houseful of gloom.

It took 2 weeks before I finally regained my strength. It was so frustrating to feel so weak all the time. When I finally felt well enough to run, I was unable to do so because I had been ordered to take it easy for at least a month. That was very difficult. Running helps me to process and resolve issues, and I couldn't do it when I needed it most.

The first few weeks were the hardest. My doctor didn't really answer my questions, and in some cases she actually gave me information that contradicted what I had read on the Internet. And I had read *everything* on the Internet! I desperately wanted and needed information and reassurance that everything would be okay—information that I could only get by waiting—but I searched obsessively on the Internet nonetheless.

The first HCG check was absolutely devastating. It was so hard to go back to the obstetrician's office and to sit with all the pregnant women in the waiting room. It didn't help that the nurse hadn't read my chart and told me to give her a urine sample. I fought back tears and could only just say "no, I don't need to anymore." Not getting it, she replied, "Of course you do. You're pregnant." I think she finally got clued in when I broke down into tears. I finally switched doctors, not because of the nurse, but because of the complete lack of confidence I now had in my doctor. There were too many things unexplained: the loss of blood, the unanswered questions, the incorrect information. I was very fortunate to find a doctor that specialized in molar pregnancies and even luckier yet, he was part of my insurance plan. I called and was able to get an appointment right away. Imagine—an office staff that was professional, unbelievably kind, and even understood what a molar pregnancy was and why I needed to come in as soon as possible to get my levels checked. It was wonderful, aside from the fact that it was on the 21st floor. I had avoided elevators neurotically after getting stuck on one with my son strapped to my chest in a carrier. We were on our way to visit my mother in the hospital after a stroke—a very stressful time for me—when we got stuck on a crowded, hot elevator. Granted, it wasn't for long, but it was enough to make me never want to get on another elevator again. And now I had to go visit a doctor and make weekly journeys to the 21st floor of a building! I was as horrified by this as I was by the molar pregnancy. I had to do it, though. My alternative was an incompetent doctor that made my heart race at the thought that my life was in her hands. So I suppose I have this whole molar pregnancy experience to thank for overcoming my irrational fear of elevators. I still don't enjoy them, but they no longer make me feel like I can't breathe.

The new doctor was great. He had answers for me. He told me that he'd never had a case in all of his years of studying moles in which a woman's levels went back up after coming down on their own. In other words, if my levels dropped on their own, the mole had resolved itself. Of course, there was a 2% chance it could happen again, but it would be unrelated to the first molar pregnancy. He also put me on birth control pills, which I had never taken before.

Like the elevator, the pills caused major trauma. Being a complete headcase, I read through the little pamphlet that came with the pills. It described how some women get life-threatening blood clots from the pills. I forget the statistics, but I had a greater chance of getting a fatal blood clot from birth control pills than I had of having a molar pregnancy. And I had had a molar pregnancy, so there was no way I wanted those pills. At the time, though, I was terrified of the cancer risk and, after much gnashing of teeth and crying to my poor husband, whose wife had suddenly turned into a hormonal, emotional wreck, I decided to take the pills.

The first couple of days I was on the pills I felt oddly euphoric. It was a very strange sensation but also a rather pleasant one. I remember smiling at the sky one afternoon, thinking how nice birth control pills were. By day 4, I felt like I was on a roller coaster. My stomach would start lurching and flipping nervously for absolutely no reason. By day 7, my hands were shaking and I was an absolute wreck. I called the doctor, and talking a million miles an hour, I described how desperately awful the pills were. I was terrified of what he might say or tell me to take in their place. His wonderful response: "Haven't you ever taken birth control pills before?" When I replied in the negative, he said, "so don't take them." I asked if it was necessary for my health to take them and he replied, "if you don't want to take them, don't take them." Sweet Jesus! Just one more reason why I liked this doctor so much.

Slowly, my numbers dropped. They were 64,000 at the time of the bloody D&C. At the first weekly test they had dropped to 13,000. Then 1,500. Then 32. Then 14. Then 9. Then 5.1. Then 2.9. Then 1.8. Then less than 1 (negative!) Then I had to get two more weekly negative readings in a row, and then nothing for 3 months and 6 months after.

It was very scary until I hit negative. At first, it was hard to deal with the loss of the baby. But I got over that faster than I thought I would, because obsessing about the details and risks of a molar pregnancy consumed me. At all times, though, it was really hard to see pregnant people or to hear about someone who was expecting. My siblings stopped harassing me about having another baby, but some friends that knew what had happened still occasionally made such stupid comments. "You really should have another one," they liked to say. Then there were all the women at the park. Our kids would end up playing together, and we'd end up talking to each other. "Is he your only one?" they would always ask. Whether it was meant to be judgmental or not, it always pierced my heart. They had no idea how that question pained me. Well, it still does, but at least now I'm free to try again.

My 3-month test was at the exact same time as my due date. I thought I was over it all and then when my due date rolled around, the loss was suddenly as raw and fresh as the day I learned I was pregnant with a snowstorm. It also happened to be the day dear hubby arrived home from work with his friend, after they had stopped out for a couple of drinks. My husband could read the distress on my face and asked what was wrong. I replied acidly, "don't you know what date this is?" Then he announced to his friend, "She has to go for her mole test tomorrow." He said this because 1) He was an idiot and thought the date I was referring to was that of my HCG test. It never occurred to him that it was also the due date; 2) He thought his drunken moron friend would provide me with some sympathetic words; and 3) He was a complete fool for never even thinking about due dates! I was floored that he just didn't get it.

So now I was miserable *and* angry, but it got worse. My husband's friend came up to me and said, "You've got to get your mole checked. You mean like this?" and he stuck a gross, fat, hairy mole on his arm in my face. I was horrified beyond belief at the time. Now I can't help but to think it is humorous that somebody could be so completely and totally thoughtless and unaware. Did I mention that this person knew what had happened and should have known better?

After that stressful 3-month period passed, life resumed. The next 3 months went by surprisingly quickly. Before I knew it, it was time for my last check. A few weeks prior to the check, I had begun to obsessively worry about what could happen if I tried this again. Miscarriages, transfusions, chemotherapy, and so on. I had convinced myself that I wasn't ready to try again and I was going to wait until I felt better about the whole thing. Then I got my last blood test. I had tears in my eyes as I rode the elevator down and walked out the door. I couldn't believe how happy I felt that it was all over! I just hadn't expected to be literally flooded with feelings of relief and elation. And I wanted to be pregnant again. I wanted to try again. That last check replaced all the misery, fear, and despair with something that I hadn't felt in a long time—hope.

It's been a full year now since I conceived what turned out to be a mole. (I still shake my head to think that I was pregnant with a mole. I mean, what the hell?) Despite the loss, I feel I've really gained from the experience. I've learned patience. I've learned to accept. I've rediscovered faith (coincidentally during the aftermath of the mole, not because of it. I came to realize how arrogant it was of me to think that there isn't a whole lot more to this world than what we can directly experience at the present.) I've also gained the ability to ask frank questions of doctors without embarrassment. And I realized that my dear hubby just can't help it if he sometimes needs a little more help figuring stuff out. Sure, I'm terrified this could all happen again. But rather than dwell on that, I've moved on. I can now be excited by a future pregnancy.

Mary

I'm 33 years old and the mother of a beautiful little girl, but we wanted to have a second child to share our love and grow our family. In 2005, my husband and I tried and tried and finally after 6 months were happy to conceive. The due date was sometime in January of 2006; however, I noticed early on (approximately 4 weeks) that I had severe nausea, bleeding, and cramping. I called my obstetrician, who asked to see me immediately. They did an in-office sonogram, but the results were inconclusive. The doctor decided to refer me to a radiologist, who found that the pregnancy was only 4 weeks along, too early to detect a heartbeat. However, there was some concern about my HCG levels being a little lower than the norm (just at 1,000), and the radiologist asked that I follow-up with another sonogram in 1 week.

I returned a week later. The results proved growth had occurred, including development of a placenta. Blood was visible in the uterus and around the egg; however, this was thought to be "implantation blood." No heartbeat was found, but the radiologist assured me that it was because the pregnancy was still young. Again, they asked that I follow-up with an appointment in a week.

I returned again, now about 6 weeks pregnant. Growth was evident, but again there was no heartbeat, and again, I was asked to follow-up with an appointment. The fourth sonogram yielded the same results. At this point, the radiologist stated that they should have found a heartbeat by now, and he referred me back to my physician. The same day of my last sonogram appointment, the physician asked me to report to the hospital for a D&C, stating that this was necessary for the type of pregnancy I had. I still had not been told that it was a molar pregnancy.

Distraught and shaking, I took off the rest of the day from work and reported to the hospital. It was only a week after the D&C that the doctor's office called me with the results. "We examined the tissue and concluded you had a molar pregnancy," the nurse said. She asked that I follow-up with my obstetrician to discuss it. Confused and concerned, I searched the Internet for molar pregnancies and read up on all the types and information so that I'd be prepared for the appointment. At the appointment, the doctor began to explain the molar pregnancy, but I interrupted to ask what type it was—partial or complete. The doctor did not know. I asked about my risks for cancer, and she said, "Less than 1%–2%." She eased my mind by stating that my type of molar pregnancy probably would not lead to cancer and that I only needed 6 months of bloodwork to test my HCG levels before I could start trying again. She added that once I had 2 consecutive months of "HCG free" status that I would be given the green light. I was very excited to hear this news and was hoping that I would get the green light in less than 6 months. However, it took the full 6 months.

Soon thereafter, I was feeling constantly fatigued and was gaining weight. I couldn't lose weight no matter how much I dieted or exercised. This was very odd for me, because I was always rather thin and very active. My weight exploded past 205 for my small frame, and I became even more distraught. I sought the help of my family physician, who did extensive bloodwork and found that my thyroid was inactive and that my blood pressure was consistently high (more than 140/90 mm Hg). I was diagnosed with hypothyroidism and hypertension, both common side effects of molar pregnancies. Since then I've been on medications and am currently taking Synthroid and Procardia.

I stayed in constant touch with my obstetrician to make sure that I was following the prescribed medications and diet for someone trying to conceive. For 6 months, my husband and I tried to conceive. Finally, a little more than 2 months ago, we had a positive pregnancy test. I was so elated. However, my obstetrician had some concerns and had me visit the office for an early sonogram. Again, the in-office sonogram didn't yield much information, and she thought it could be due to her equipment or that the pregnancy was just too young.

Three days later, she sent me off to the hospital for another sonogram. The results yielded a 5½-week-old fetus with a heartbeat of 105 bpm. Again, the doctor was concerned, stating that 110 bpm is what they normally look for. However, I chalked up the doctor's concern to paranoia. I kept telling myself that the doctors were overly concerned because of my history and that the chances of another loss were less than 1%. My logical mind was convinced that statistics were on my side.

A week after the hospital sonogram, the radiologist zoomed in on the baby. I had a tinge of pain as she pushed the instrument up inside me—something I had not felt before with sonograms—but I dismissed it. As she zoomed in, I could see the fetus. I saw a head and a heartbeat beating rapidly. The equipment captured a heartbeat of 130 bpm. The radiologist even smiled and said, "Congratulations. You are going to be a mommy, probably in early January." I was elated and had new confidence. I was totally convinced that the doctors were paranoid, as I had suspected. I went home to tell my husband to celebrate. Later that night, I got a call from my obstetrician. "Mary, I just wanted to congratulate you. We got the results and I had to tell you before the weekend that everything looks healthy." This gave me even more confidence that all was well, and I began to tell our family and friends and even strangers I just met.

I had a standing appointment on the calendar to see my obstetrician a week later, and she asked that I keep it given my hypertension. I showed up at the office smiling and giddy. The doctor told me that the due date was the exact same due date of my little girl. What a coincidence! She then decided to do another pelvic sonogram. Her face was cold and distant. The same face when I had the molar pregnancy. She said, "I'm a little concerned that I'm not picking up a heartbeat, but that could be due to the fact that my equipment is not as sensitive." Again, I dismissed what she said. She had me follow-up with a sonogram at a radiologist's office the same day and asked that I call in sick to work, just in case. I followed her suggestions and thought to myself, "She's too paranoid and needs to loosen up. And, how cool is this to get a day off from work, prescribed by my physician?"

The radiologist technician examined me. Again, a cold look. But hey, she was young and just a technician…. She ran to get the radiologist on staff. Again, the same cold look. They examined me over and over again. Every time they pressed the equipment up inside me, I cringed with pain. It was more uncomfortable than the sonogram from the week before. I watched and listened as they tried to pick up a heartbeat. The monitor showed nothing. But who can make out that white snow effect anyhow? The radiologist looked at me with sad eyes and said, "Mary, I'm very concerned. We are not getting a heartbeat. And for the age of the pregnancy…" I interrupted him right there. "What do you mean, no heartbeat? Your equipment must be bad! I had a heartbeat of 130 bpm last week! Test again." He replied, "That makes me even more concerned. If you had a heartbeat last week, and we aren't getting one this week...." Right then and there I lost it. "Test again!" I told them, but he replied, "I'm sorry. Our results are conclusive." They asked me to get dressed and offered to call my physician's office for me.

My physician saw me right away and broke the news that I already knew I was going to hear. She told me that the positive side was that it wasn't a molar and that I had already had one successful pregnancy. But that didn't seem all that positive. I had lost, and I still needed answers. As I write this, my D&C is scheduled for Tuesday, and right now, all I can think about is getting past this. The thought that I might spontaneously start bleeding in public frightens me. The loss plus the embarrassment and humiliation is something I don't want to endure.

Explaining the loss to friends, family, and strangers I had met was hard, but it was especially hard explaining to my child. She had the same questions I did. She asked me why the baby in my belly died. I didn't have an answer other than "sometimes this just happens." She asked me if we were going to put the baby in a small box and have a funeral and bury it. I didn't know how to respond other to say, "Yes, the doctor is going to put the baby in a small container for burial." This was partially true; my doctor is offering to do genetic testing on the tissue. My husband is all for getting answers and scientific results as to the root cause. His family is against it, however, saying that we should just let the past be the past and not relive the pain. Right now, I'm mixed up. Was my blood pressure medication—the medication that my family practitioner, obstetrician, and high-risk specialist all agreed to—the root cause? Was it my thyroid hormones being out of whack? Was there leftover molar tissue in me? Was it my age, at 33? I can't imagine it was anything else. I totally detoxed my body of caffeine, alcohol, wheat, and dairy months prior to conception. I lost 20 lb and was exercising four times a week. I was healthier than I had been in years.

What is plaguing me now is, if I try again, can I survive another loss? If I find out the cause and it is not in my control, just bad luck like the molar before it, can I take that risk again? The scientific reasoning side of me wants to know my chances, but the emotional side of me doesn't want to put myself through the pain again.

Megan

On February 9, 2002, I had the feeling that my body was changing. My husband and I went grocery shopping and as he was picking up the last few items, I snuck off to the pharmacy section and picked up a home pregnancy test. I was only a few days late, but I had been feeling a little strange and suspected pregnancy.

When we got home, I rushed upstairs and took the test. To my shock, it turned positive within seconds. We had decided to abandon birth control the month before and let things happen naturally. I never imagined that it would happen so fast! I told my husband the news, and together we absorbed the shock. We were totally astonished but happy as could be.

I went to my doctor 2 days later, and she confirmed my pregnancy. I was almost 6 weeks along. I was completely thrilled but nervous just the same. This being the first time, I wasn't sure what to expect. My husband and I went out for a nice Valentine's Day supper a few days later. We walked to the restaurant, and when we got home, I noticed some spotting. I was concerned and called the doctor the next morning. She told me that it was a common symptom for many women in early pregnancy.

The spotting came and went over the next several days. I finally called the doctor the following week, and she suggested I come to see her for an examination. I went, and after she examined me she said that the spotting was coming from my cervix and that there was nothing to be concerned about because this was normal for many women. I felt reassured, but at the same time I had a nagging feeling in the back of my head that something was very wrong. I kept telling my husband that it didn't "feel right."

On February 25, I experienced excessive spotting. I became hysterical and called my doctor, begging her to give me an ultrasound. She called the hospital and set up an appointment for the next morning. When my husband drove me to the hospital the next day, I actually hyperventilated as we approached the hospital. I just knew it was going to be a horrible experience. When they performed the ultrasound, the technician didn't say anything. I asked her if she could see the baby, and she said she had to go get the doctor on call. The two of them examined me for another 15 minutes and still nothing; 20 minutes later they informed us that they couldn't find the fetus in utero but noticed a bump on my fallopian tube and suspected I was having an ectopic pregnancy.

I was then rushed to the operating room for the first surgery of my life. I was so scared and so crushed that my baby was not to be. I woke up to my doctor telling me that it had in fact not been an ectopic but just a cyst in my tube and that they had to review my ultrasound results again to determine exactly what was going on. I went home with the hope that my baby was still in there somewhere and they had just made a mistake.

The next afternoon, the doctor called and told me that the reports showed I had had an incomplete miscarriage and I needed a D&C immediately. I was overcome. All that hope was now gone. I went in the next day for the D&C and couldn't wait for the nightmare to be over. When I woke up, they told me it looked like a typical miscarriage, but they needed to receive the pathology report before they could give me a definite answer.

Two weeks later I was informed that what I had had was a partial molar pregnancy. I was devastated.

I started my weekly testing and went from 85,000 HCG down to less than 2 in about 9 weeks. Less than 5 was considered the "zero zone" by my hospital. After three weekly tests rating at less than 5, I could start monthly testing. My first monthly test was the last week of May 2002. My doctor said that if my levels remained below 5 for the next three monthly tests, I could go off birth control and it would be okay to conceive again.

At the end of July 2002, 5 months after my molar pregnancy was diagnosed, we were given the all clear to try and conceive again. After going through the worry of many weekly and then monthly tests, it was a welcome relief to know that everything was okay. On October 31, 2002, I found out I was pregnant. We were thrilled but a little nervous, having been through such trauma only months before.

My doctor sent me for an ultrasound at 7 weeks, and there I saw my baby with a beating heart—the most magical thing I'd ever seen! Five weeks later we got to hear the heartbeat, and I went on to have a wonderful, healthy pregnancy. On July 4, 2003, our beautiful healthy son was born. He is my special gift from God and I cherish him with all that I am.

The molar pregnancy was devastating. While recovering from it I tried to keep a positive outlook, although it was very difficult at the time; some days I thought I'd never feel whole again. I will never forget that first pregnancy or the due date when our baby would have been born. My focus is now on my son—I look at my new little boy and feel so blessed to have him. He just fills my heart with such joy. He is so worth all the hardship and waiting we had to experience before we were blessed with him.

I just wanted to let people know that we did have a happy ending to our story, and I want to encourage people experiencing the aftermath of a molar pregnancy to keep going. You'll have good days and bad days, but it will get a little easier as time passes. You'll never forget, but you will be able to cope. And when you finally do get the all clear to try again, keep the faith and a positive outlook. My heart goes out to each of you.

I also wanted to say a big "Thank You" to my husband and family, who were such a great support. I know I couldn't have made it through without their love and encouragement.

Meghan

In the fall after my college graduation, when I was 22 years old, something strange happened to me. Although I am an only child who had always found the presence of children somewhat disturbing and had never even held a baby, I found myself afflicted by "baby fever." I was a bit perplexed as to where exactly this was coming from. I was engaged, but the relationship was becoming strained. In November 2002, during a short trip to a wedding in Mexico, I pondered my strange new appetite, the tears I shed at a *Charlie's Angels* movie (and me normally such a stoic!), and after seeing a baby sleeper, and my feeling just like MUSH inside. Sure enough, my period was delayed, and sure enough, the two lines in the pregnancy test I took during my lunch break at work were too bright to be ignored.

Thus began my evolution. I was becoming a new person, a person I never thought I could be. I was a mother. I felt it radiating through every cell in my body. Although I was filled with happiness and love and power and energizing, life-creating creativity, I was also stricken with an undeniable form of terror. It was so terrifying that I could be given something that changed my very essence and that it could just as easily be taken away. I am a natural-born pessimist, and in the back of my mind, I was always thinking of miscarriage. At my very first ultrasound, I was terrified that there would be nothing, no heartbeat—nothing. However, when the sound of rapid pounding radiated through the air, I thought to myself, "Thank God. I can let go of some of this worry. The heartbeat is the first step to a baby 8 months from now!" The ultrasound technician noticed a small blurb on the screen aside from the baby (who I affectionately called Gummy Bear due to the little stubby limbs) and suspected either a fibroid or a twin! Although thrilled at the latter possibility, later appointments with my midwife confirmed that the mysterious shadow had somehow drifted away. When I think about that moment in retrospect, that shadow truly was a cloud of darkness, ominous and threatening, and it would bring destruction in the end.

Things were really not going well with my fiancé, who I married in something of a shotgun wedding due to the pregnancy (cultural constraints led us to marriage). We were keeping everything a secret for the most part. We had not told any family members, even 2 months into the pregnancy. He was truly being awful. He was gone day and night, only he knows where, usually coming home smelling of alcohol. He caused me so much distress, I preferred not to think of him at all. It was MY baby—it was almost as though he had no part of anything. Perhaps I am just remembering it this way now. It is hard of me to think of him fondly when remembering my pregnancy. I only recall anger, pain, and suffering. Only a few months into the pregnancy, I was already planning on MY life with MY child—a life in which he would play no part.

Things were going well. Although I was frequently nauseous, I rarely vomited. I had all the normal pregnancy quirks: super sense of smell, weird cravings. I planned to name my baby (whom I was convinced was a boy) Cesar. I couldn't wait to pick out all those tiny clothes and shoes and supplies. Everything was carrying along normally, and I was truly basking in the glow of being pregnant, in a way I didn't even think was possible for me. I was surprising myself!

I went for an appointment with my midwife every month. During the third month, when I was about 16 weeks or so, the midwife noted that my blood pressure was a bit elevated. The baby's heartbeat was normal. I was fine. It was just that the spike in blood pressure was a bit of cause for concern. She advised me to purchase my own machine and to record my pressures. I also made an appointment for an internist for the following week.

In about a week, things began to deteriorate with my health. It happened so rapidly that it is hard to pinpoint now how everything transpired. Although I was still working and carrying on as usual, everything was a thousand times more difficult than it had been. Walking from my counter at work to the bathroom forced me to summon every ounce of energy I had and then some. My entire body was painfully swollen. I was vomiting five or six times a day. Anything that went in came back up instantly. At one point I vomited pure acid. It was just terrible. I was recording my pressures, and they seemed to get worse and worse as the days went on. I had headaches so painful I thought my head would explode. At one point I stared up at the ceiling in the dark and everything went black as I passed out from the pain. My dog was pacing and barking and the hair was standing up on the back of his neck like there was a ghost in the room, and I felt that there was.

Finally, when I was about 17 weeks, I went into the bathroom at night and noticed a smear of bright red blood. My heart fell out of my chest. Although I thought that because I had made it to the second trimester I was "safe," that blood filled me with dread. The midwife advised me to go to the women's hospital. When I did, my pressures were still extremely hypertensive, but they were reluctant to diagnose me with pre-eclampsia because of the extremely early onset. My husband was with me, and he heard the heartbeat for the first time (he had never been with me before at an appointment), and I heard it for the last. It was a Thursday night. I was given some kind of migraine medication, told to make another appointment with the internist (who had by that time put me on beta blockers for the blood pressure, although they were not working and I even had to double the dosage), and sent home. I vomited four times on the way home, but I had heard the heartbeat and that's all that mattered.

The following Thursday I truly felt like a shell of a person. Looking back, I cannot believe that I was working full time, going to graduate school, and doing everything else as though nothing was wrong. Although I was about 18 weeks pregnant, I wasn't showing yet. I felt like a walking corpse. I called my mother from the grocery store at about 11 P.M. at night and told her my pressures were still averaging 200/140 or so. I had been to the internist that day, and he said I had to go on bed rest, no question about it. I also had proteins in my urine and was getting ready to take a 24-hour urinalysis. However, my mother demanded I go to the emergency room because I could have a stroke with those pressures. Downplaying things as usual, I refused. She is a nurse at a very far suburban hospital, and she was on shift at the time. She said she would meet me at the front door and walk with me into the emergency room and help me. I had just told her about 2 weeks before that I was pregnant and was surprised to discover that she was as thrilled as I was! Never mind that it wasn't my normal hospital and that I had been to the doctor earlier that day. She wanted me there.

My husband and I made the 45-minute drive, and sure enough, my mother was waiting for me in the doorway. I went to her, and my husband drove off. No surprises there. Once in the emergency room, which was silent due to the far location, the nurses and doctors were shocked to see the pitting edema that consumed my lower body. I looked at my ankle and it was so swollen I thought it would burst. The urine test confirmed that I was dropping proteins at 4+, a very high number! I needed to do an ultrasound, but first I needed intravenous fluids. Getting anything intravenously is my biggest phobia, so it took about an hour just for a nurse to coax me into doing it. Oddly enough, at this point, I was really only worried about that; I was denying anything else was wrong. I went in for my ultrasound. Although the face of the technician was somewhat grim, we chatted casually about my plans for my baby, my school plans, anything. She kept saying over and over "I can't make a diagnosis, the doctor will be reviewing the slides." I hadn't even asked for that information, she just felt the need to keep telling me and telling me.

When the nurse attempted to get a heartbeat with the fetalscope and was unsuccessful, she tapped it against the counter and said reassuringly, "Oh, well, you know how these things are. Sometimes the baby hides or is positioned funny. And of course, these fetalscopes are always breaking down!" I ate it up. I knew it was a lie, but I ate it up.

The emergency doctor, a kind, older, bearded man, did not have such a cheery attitude. He HAD reviewed the ultrasound slides, and he had nothing funny to report. It is hard for me to even type this. I can see everything in my mind perfectly, even almost 4 years later. He told me so matter-of-factly of the "fetal demise." I don't remember if that is specifically what he said, but I prefer to write it that way now. The typical grape-like placenta had been seen on the ultrasound, and that bounding heartbeat I had heard a week before had quietly extinguished. No wonder I felt like death and decay. No wonder I felt supernatural. No wonder my dog had been battling demons and ghosts in the night, viciously attacking the air. They had come and taken my baby away.

In the morning I was flooded with information. I was given two choices—delivery or D&E. I bawled and bawled. I felt that I had retched out all of my insides with my tears. I had phoned my husband from the bathroom, but he was in a bar, and his phone was off. I all could do was leave a message to tell him that the baby had died. He arrived about an hour later, so shocked (and most likely guilty) that the alcohol wasn't even having an effect on him, and he suddenly morphed back into the person he had been before. I was terrified of going into labor and delivering a clotted mass (along with my tiny dead baby) and also felt I would be blaspheming him to have him ripped out and sucked apart during the surgery. I didn't want to take him out of me. I wanted him with me! When presented with the two choices, I really just wanted to die. It didn't matter if I died, as long as the baby was with me. My mother finally convinced me to do the surgery. At least I did not have to bear witness to it, and I was comforted by the fact that I could have my baby cremated and could hold onto him forever.

I hated the obstetrician that attended my surgery. He referred to my baby as the "products of conception" and stated that the D&E was a process in which "the products of conception would be suctioned." I wanted to strike him, hurt him. I hated him; I hate him even at this moment. The internist also phoned that same morning, leaving a frantic message on my phone. My lab results had come back. Everything was dangerously off-kilter (particularly the thyroid), and the pre-eclampsia was very severe. He wanted me to admit myself to the hospital immediately. I wish I had been there, at my own hospital, where perhaps I would have been treated with some degree of respect.

Just before the surgery, I had a mental breakdown. It was a combination of my medical phobias and overwhelming grief. All I remember is being sedated. A river of warmth ran through my body and hit my head with a passionate force, and when I woke up I was choking to death. I tried to call out to the nurses I could see in the corner of my eye, but I was voiceless. I tried to raise my arms to grasp my neck, but it was futile. I was going to die right there. I closed my eyes, submissive to death. A tube was ripped from my throat and I was wheeled down a hallway. I slept and slept. When I did wake up, I saw blood everywhere. It was to be expected—my life had poured out of me. Everything felt like blood and death and pain. Nothing was real. I was trapped in hell. There was some kind of sign on our door, with doves and hands (signifying loss), and I hated it so much I wanted to rip it off.

The doctors talked on and on about weekly blood tests, HCG levels, possible cancer, partial molar, complete molar, pre-eclampsia, methotrexate, metastasizing, and so on. I was thirsty for the technical knowledge, but none of that remains with me now. I remember the loss more than anything. I remember the following few days in the house, mostly alone. My mother-in-law stayed with me for a time. She watched me as I crashed from the hormones, as I panicked, and wept, and screamed aloud for my baby. I pictured a neat little bundle—one I would never hold. I can't even imagine feeling that way now, I have removed myself so far, become so anaesthetized to it. She told me of a dream she had had the previous week. She hadn't even known that I was pregnant. Yet she dreamed of a little tiny blanched baby, no bigger than the palm of her hand. It came to her, and she knew it was an angel. She just held it and wept in the night. She and I connect on so many levels, so deeply, and I was so happy that she too had known my baby. I even clung to my husband, knowing it would probably be the last time. I knew he would revert back to his ways, but for the time being I needed him.

Testing the blood levels each week was a very welcome distraction. I still got to go to the clinics like a normal pregnant woman and have tests. Everyone that saw me in there knew I had been pregnant. I knew a time would come when no one would ever think of me as ever having been pregnant (and of course, that time did come), and my near-status as motherhood initiate would be stripped from me. So the levels gave me one last chance for my pregnancy to endure. In a way, it really was enduring because I was still filled with the hormone that had helped to cultivate my baby. They started out near 1 million. The following week they had dropped to about 10,000. Two weeks after that they were at 1,000, then 100, then 50, then 15, then nothing. The grief did not lessen nearly as neatly as the HCG levels. It ebbed and flowed. Some days it was only in the back of my mind, and other days it consumed me. I named the baby Plantagenet, after a medieval royal dynasty (which is only fitting since I was attempting a Master's degree in English history) and had his name engraved on a ring with two birthstones, one for his conception month, October, and one for his death month, February. I pictured him in the womb, fighting like a royal medieval prince in full armor, charging valiantly against the tumor that threatened to consume him. My mind flashed back to the first ultrasound when the shadow "twin" had been detected. It was that which would gradually become the molar.

The next year or so I went in and out of myself. I would be at work and hear a pregnant woman carrying on about her baby and break down. When people asked me if I had children I would say "Yes, but not on earth." I struggled to hold on to my status, even though it was becoming real to me that I could not, in fact, claim to be a mother any longer. When I would try to forget, Plantagenet would force me to remember. Still to this day I feel we have a celestial connection of sorts. Although I have a very busy life nowadays and am single, when I feel the "baby fever" I wonder why until I realize the date. I am always filled with feeling for Plantagenet near his due date (July 12), his time of conception, and the time of his death. On these days, I take to wearing the ring again, begin to remember ever so slightly, and light a candle for him—one that fills the entire room with the scent of roses.

Truly I am not sure how this has all affected me. I can only hope to examine my subconscious to find out. I have reverted back to my old self, even worse than I had been before, shunning the company of children and babies, thinking myself entirely inadequate to care for them, not caring for their company anyway. I have not spoken to my ex-husband in about a year. The thing that pains me the most is that he hid the ashes from me when I left. I don't really need them anymore, however, because ashes are scattered all over my heart. There is a black hole there. I have a boyfriend now, but even with him I feel myself to be very closed off. He has never known that girl, that mother. Something shut down inside me from this experience. Of course, I have plodded onward, and to know me and speak to me one would NEVER imagine that I have endured this. I do not walk around in some sort of noticeable state of depression; I am happy and motivated and successful with the best of them. Sometimes, I don't even feel like I did endure this. That person, that MOTHER, seems like someone quite apart from myself. But sometimes that baby will enter into my thoughts and remind me that that mother is and will forever be ME. That person is inside of me, hidden from the world, just as my baby was during his time on earth. No one will ever know my baby but me, just as no one will ever know the mother that I was for a brief 4 months those many years ago, at least not for the time being.

Melissa

My name is Melissa, and at the time of my partial molar pregnancy, December 2003 through April 2004, I was 30 years old. I had two little boys ages 4 and 2 and a stepson who was 10 years old.

When we learned we were pregnant with baby number three we were really happy; we were a little antsy about maybe having another boy, but happy nonetheless. Things started off as they always did for me, with throwing up. We had a confirmation ultrasound at 6 weeks, and our little peanut's heart was beating strong.

At 8 weeks I had my first official obstetric visit. We heard the baby's heartbeat again, faster than the other boys' had been, so maybe it was a girl? Everything was fine, I was really sick, but again this was par for the course. Things ticked along nicely. I craved a lot of Mexican food, which of course I indulged. The boys kept me really busy, and before long we were at 12 weeks.

The next visit to the obstetrician involved all the usual questions. Yes, I was still very sick. The doctor prescribed some medicine for the morning, noon, and night sickness. We heard the baby's heartbeat, still strong and fast, and kept our fingers crossed for a girl. We scheduled bloodwork for the quad screening. I'd had mixed results in the past with a false positive with my first son.

At home we were busy preparing for my father to visit from Australia. He usually comes when the babies are born, and things are chaotic, so this time we decided maybe he should come mid-pregnancy, while I'm still mobile, and then again later when the baby was settled. At 16 weeks I went in for the usual checkup. I remember the day because it was April 1, April Fool's Day. Things were hectic; I had my 2-year-old with me and was fitting my appointment in while my 4-year-old was at preschool. The nurse took my blood for the quad screen and tried as usual to find the baby's heartbeat using the Doppler. She tried several times but found nothing. When the obstetrician came in she had a portable ultrasound machine. She made light of the situation, saying it was probably nothing, just my little peanut hiding. She found the baby with no trouble, and the heartbeat was slower, about 120 bpm. She thought the baby was probably resting. I hurried back to the preschool and then home to get organized, because my Dad was arriving in 2 days.

On the Sunday evening I picked up my dad from the airport. He had never seen me pregnant before, and he said it suited me. He said I was glowing, something I've always thought sounded silly. I was grateful, though. I had finally stopped throwing up constantly, so maybe that was it.

On Monday the obstetrician's office called. My lab results were in, and there were some "concerning" results. I needed to schedule an appointment for a level 2 ultrasound and to see a genetic counselor. I was concerned but was also a little dismissive; we'd been through all this before with our 4-year-old. We had even had an amniocentesis, and everything turned out perfect that time. It wasn't a big deal now; maybe I'd get to find out the sex of the baby early. I couldn't get an appointment until Thursday, however, and as time went on I grew more anxious. The nurse hadn't given me specifics on what was wrong. Thursday took what seemed like forever to come.

Finally it was Thursday. My mother-in-law wanted to come with us to get the early news on the sex of the baby. My friends watched the boys so that there were no distractions. It was late in the afternoon, so there was no parking at the hospital, and the doctors were running late. Finally we got in to see the genetic counselor. A lovely lady, she went over the results. I'd tested positive for trisomy 21 and 18. Down syndrome? I'm only 30! The counselor stressed that it could be a false positive. We were reassured and thought, here we go again!

We went in to see the doctor then, and she remembered us from my oldest son's ultrasound and amniocentesis. I felt comfortable. As she started the ultrasound I looked at the baby, straining to see the sex, and asked what exactly we were looking for. She gave no response, but just kept running the Doppler over my belly. Then she said, "I'm sorry to tell you, your baby has died." What? I'd just heard the heartbeat on Friday! She then left us alone and asked if in a little while we could resume the ultrasound if I felt "up to it." She wanted to make some diagnostic observations and have a colleague take a look. Take a look? I was stunned. My husband was crying and hugging his mom. I was laying there looking at my tiny dead baby. All I could think was where did this go wrong? Sometime between Friday and Thursday my baby had died. I didn't even know, shouldn't I have felt something? The doctor returned with another specialist and they looked at the ultrasound, printed some pictures for us, and told us that it was "classic" triploidy. She also said I would need to have a D&C, and she thought it was better if I had some time to grieve before the procedure. All I wanted was to hold my children.

My obstetrician's office called later that day to schedule an appointment for the next day. My mother-in-law graciously offered to watch the boys. That night was the longest night of my life. I was devastated and angry. How could I not know? The next day was Good Friday. My appointment that day was not for the D&C, as I had thought, but rather was for a consultation to discuss the situation. How long was I expected to walk around with my dead child inside of me? The doctor said she was booked solid and couldn't possibly "fit me in" that day. I would have to wait until Monday—2 more days. Emotionally and psychologically I wanted it over. I called the hospital and told them that I didn't care who did the procedure, that it needed to be done, and it could not wait. They fit me in that night, with—surprisingly—my own obstetrician.

The D&C was uneventful; I was ashamed to be there. How could I do this? I had no option—the baby was dead; yet I felt such guilt. As I changed into my gown I said good-bye to my tiny little baby. I told "her" how very sorry I was for this happening to her and that I loved her very much, always. Coming out of the anesthetic while I was being wheeled out of the operating room, I remember asking my doctor if it was a boy or girl. Later I was embarrassed. I don't know why I thought things were okay; they were never going to be okay. I went home 2 hours later.

The obstetrician's office called the next morning. Results indicated the pregnancy was "classic" triploidy, but there would be further testing that could take up to 2 weeks. I was scheduled for a chest X-ray and bloodwork the same day. I was confused about the need for the X-ray; did they know something I didn't? The bloodwork, they explained, was to ensure my hormone levels were returning to normal. I was told to return every week for 6 months. Nothing made sense.

My doctor's office called about a week after my D&C to tell me that preliminary results indicated that I had carried a partial molar pregnancy and that I needed to schedule an appointment to discuss this. When would this end? My doctor is young, maybe 35, and has no children of her own. She opened the textbook in front of me and read my condition to me. She explained about the causes, which were largely unknown. She explained about hormone levels and the need for ongoing blood tests, and then explained about not getting pregnant again because this horrible situation—this thing that had happened to my baby—could develop into cancer. What horrible thing did I do to bring this upon myself?

Week after week I got my blood drawn and tested. My HCG dropped to 6 within a week, then stayed there for a month. My obstetrician was not concerned, because it was "dropping nicely" and not rising. I was concerned, however. Not only had I lost a baby, but my children could lose their mother! Finally the level dropped to what I now know is a "trace" levels. Then began the wait; I felt like I had been given a sentence. I absolutely wanted another baby.

My husband and I went through a lot. He wanted to forget what had happened and not talk about anything. I wanted to remember our little baby, who we finally found out had been a little girl. In September I was given the "all clear"; 6 months had passed and my condition had not changed. My obstetrician still did not condone having another pregnancy, however; she said "guidelines" preferred a year wait. If I did become pregnant, though, she wanted me to come in as soon as I knew.

September 20 came and went like any other day. How was anyone to know it was my baby's birthday? My husband bought me a potted garden, which was interesting because I have a bad track record for keeping plants alive. What if this one died too? What would I have then? I put aside all the feelings this plant brought up; it was a gesture and that was all.

My husband and I talked and argued, and I think I even begged, about having another child. September came and went with no pregnancy; I became convinced I had developed secondary infertility. I had gotten pregnant so easily the past three times, what had this ordeal done to my body? What if I couldn't have another baby? My husband was neither enthusiastic nor encouraging about having another pregnancy, nor was he sympathetic when it didn't happen.

My mum came to visit from Australia in October amid this conflict. She understood the need to "make things better" for me, but she also knew my husband was afraid of a recurrence. It was a hard time for her to be around us, I'm sure. She flew out on Halloween, but not before interfering a bit on my behalf, which of course earned her no points with my husband. This made things between him and me more difficult for a while because I had, as he put it, "called in recruits." I appreciated her support nonetheless; it was nice to feel like I wasn't going it alone for a change.

In mid-November I went in for a monthly blood draw, and nurse called me back the same afternoon. I was pregnant. My instinct was to call my husband. How could he not be delighted, overjoyed, relieved—I'm not broken! Well he wasn't. He wasn't even happy. His exact words were "are you satisfied now? You got what you wanted." I was very satisfied! Happy, scared, the whole gamut.

We took every test know to man. I had a confirmation ultrasound at 6 weeks, and everything "looked good," but having been down that road before I decided not to tell anyone else. Not until we knew for sure that this pregnancy was okay! After having another ultrasound and a test known as a nuchal translucency screening at 12 weeks and getting good results from both, I was ready to tell family. Yet the reactions were mixed. Why could no one be happy for me? At 16 weeks I had another ultrasound and the quad screen—the dreaded quad screen—and got the incredible news I'd waited almost a year to hear. We were having a healthy baby girl. I was offered an amniocentesis but declined; I couldn't take any risks this time. Everything had come back clear; there was no reason to suspect this baby wasn't perfect. We scheduled another ultrasound for 24 weeks and another for 32 weeks, just to "keep an eye on her." They weren't medically necessary, but my obstetrician indulged me.

I was scheduled for induction on my due date, August 5, 2005. The summer dragged on and on. The day finally arrived, but I was bumped because my induction was considered elective. I was furious. I don't think the hospital had encountered such an angry pregnant woman before. The induction was rescheduled for the following Tuesday, 4 days later. Four long days! I needed so badly to hold her in my arms, how I waited for her! No one understood my frustration or impatience, they kept saying, "it has to happen eventually."

Everything went smoothly, the induction, the labor, and the birth. Sydney Annalise was born on August 8, 2005, at 2 P.M. When I saw her I wept over my tiny, squirming little baby. My heart ached. Incidentally, she wasn't all that tiny. She was 8 lb, 5 oz, and 20 inches long! She turned 14 months this week, and she's walking and talking and getting into all kinds of mischief. She bullies her brothers and torments the dog. She is totally spoiled and totally adored, even by my husband. When I look at her my heart aches. I hold her so tight sometimes and think of what I lost, but then she kisses me (or bites me, as it's been lately), and I'm reminded of what I've gained.

Michelle

My husband, Dimitri, and I got married in August 2002. We found out we were pregnant just a few months later, in December 2002, just before Christmas. Although we were not trying to get pregnant, we were not preventing it either. We were in the mindset that if happened that would be great, and if not, we had plenty of time. I remember taking the pregnancy test and being so shocked and excited! We told all our family, who told everyone else; in a short amount of time it seemed everyone knew of our exciting news. I went to the doctor for my first visit at 7 weeks. The doctor did a routine ultrasound, and we saw a baby with a heartbeat. The doctor did tell us that the baby didn't look as big as it should, but she said not to be worried. She wanted me to come back in a couple of weeks to check the baby again. It never crossed my mind that anything could be wrong. I was so happy to be pregnant, and I *felt* pregnant. I felt bloated and had morning sickness. I thought everything was fine.

At 9 weeks I went back for another ultrasound. This time we saw the baby, but there was no heartbeat. The news was shocking. I was stunned. I had had no spotting or cramping. I thought everything was fine. Everything after that seems a bit fuzzy. The doctor told me that I could wait and miscarry on my own or have a D&C to remove the pregnancy. I chose to have the D&C. I thought the sooner this pregnancy was over, the sooner we could try again for another baby. Of course during this time I cried. I spent a few days on the couch just crying and crying. Why was this happening to me? What did I do wrong? What I didn't know was that the news would get worse.

A week after my D&C I got a call from my doctor, who told me that the results from the D&C showed I had had a partial molar pregnancy. What the heck was that? The doctor tried to explain it to me, but again I was in shock and the conversation with her seemed fuzzy. I heard that I would need weekly blood draws and that if my HCG number didn't fall to zero that would mean I would need chemotherapy. She also said she wanted me to wait a year to get pregnant again. I called my mother, who told me just to do what the doctor said and everything would be fine. My husband during all of this was just as shocked and saddened as I was. He didn't know how to help me. It must have been frustrating for him. I, of course, went on the Internet and starting researching, trying to find every bit of information about molar pregnancies that I could.

I also had to return to work. By this time the news had spread that I had lost the baby. Some people were very nice. They offered support, said they were sorry, or sent cards. Other people talked to me as though nothing ever happened. Some people would tell me it was nature's way of taking care of things, or that I was young and we could try again. All of these responses made me angry. It wasn't fair! This wasn't supposed to happen to me! I tried to tell some people that this was not just a normal miscarriage. That I had to go through weekly blood draws. That if my HCG levels don't go back to normal I would have to have chemo. That I was suppose to wait a year to get pregnant again. When I would tell people this they usually looked at me with a blank face, not knowing what to say, not knowing what I was talking about. They had never heard of a molar pregnancy. I felt so alone and isolated. My family was there to lean on, and my husband offered lots of support, but I never felt as though anyone understood.

So I started on my weekly blood draws. I didn't write down the exact numbers, but I do know my HCG levels started to fall right after the D&C. I started to have weekly blood draws in February 2003, and by April 2003 my HCG levels were at zero. What a relief I felt. No chemo! But the doctor still wanted me to wait a year from the point my levels fell to zero to get pregnant again. That would mean I couldn't get pregnant until April 2004! In the meantime, it seemed everyone I knew was pregnant—several girls at my work, friends, my sister-in-law, everyone! I think at one point I knew of 10 pregnant women. Everywhere I went people were talking about babies and pregnancy. It drove me completely nuts. Everyone thought that I should be over my loss, and they didn't understand why I didn't want to come to their baby showers, or why I didn't want to hear about their pregnancies. My heart ached. Why couldn't someone understand my pain? Why couldn't this have happened to someone else? I cried every day.

I finally decided that I needed to go talk to someone about this. I made an appointment with a counselor. She was a nice lady who understood that having a miscarriage was hard. But she never understood the molar pregnancy thing. She never understood that I was scared that this would lead to cancer, or that I was so frustrated because I was told to wait to get pregnant again. After a few sessions, I stopped going. Again my husband was always there to support me and to let me cry on his shoulder, but he didn't know what to do for me. I didn't know what to do for myself. I knew the only thing that was going to make me happy again was to be pregnant.

I remembered that in those few days after the diagnosis I had read about online support groups for people who had gone through a molar pregnancy. I signed up for one, and I began to realize that I wasn't alone! There were women out there who knew what I was going through. They understood my pain and sorrow, my need to get pregnant again. This was exactly what I needed. I learned that for a partial molar pregnancy many doctors were now saying that a 12-month wait was not necessary and that 6 months was plenty of time. I marched into my doctor's office and told her that in September (6 months after my HCG fell to zero) my husband and I were going to try to get pregnant again. She advised me to wait the year out. She said I was young (26 years old) and had plenty of time, that there was no need to rush. I told her I didn't care how much time I had. I was determined to get pregnant again.

Just a month later, I was back in the doctor's office ready for an ultrasound. Again, I was 7 weeks along. I was so nervous during the ultrasound, I couldn't stop shaking. But we saw a baby with a heartbeat. The doctor said the baby looked great! She congratulated us and told us just to be cautious, that I should come back in 2 weeks so she could check the baby again, but she saw no problems. I left the appointment beaming! I was so happy. My husband I had learned a lesson from the first pregnancy, however; this time we only told our parents and no one else.

Two weeks went by, and we were back in the doctor's office for another appointment. I felt nervous, but I was sure that everything was fine. All my fears came true during the ultrasound when I saw the doctor's face. She told me that there was no heartbeat. I had again lost the baby. Now that I look back, I should have realized something was wrong because I never felt any pregnancy symptoms. The doctor this time insisted that I have a D&C to make sure this was not another molar pregnancy.

Again, I was shocked. I cried and cried and cried. Why? Why? I kept asking. Was I never to have a baby? Why could everyone else get pregnant and I couldn't? What was wrong with me? Maybe I should have waited the full year? Again, my heart ached.

I had my second D&C the day before Halloween in October 2003. The test results came back that it was not a molar pregnancy! Thank God! It was a normal miscarriage. I wouldn't have ever imagined that hearing I had a normal miscarriage would be good news. Of course I was sad. But it was a bit easier to deal with this time around. Not many people had known that I was pregnant, so I didn't have to deal with people saying the wrong thing, or not saying anything at all. I also had my online molar support group, who offered me the support I needed. The doctor put me back on weekly blood draws, but my HCG levels fell to zero immediately. The doctor told me to wait 3 months to get pregnant again. She said my body had been through a lot and that it needed to rest. I figured I could wait that long.

December came, and so did the flu. It was the year that everyone had the flu and there was a shortage of flu shots. I was so sick I couldn't leave the couch for 10 days. It was the type of sick when your body aches, your stomach churns, and you wish had gotten the flu shot! I soon recovered from the flu and starting thinking about getting pregnant again. I knew that doctor had told me to wait 3 months before trying again, but I was preparing myself. I was watching my cycle so that I knew exactly when I ovulated and my husband and I would be ready. Little did I know that having the flu for 10 days could make ovulation come late.

I told my husband not to worry about protection, I had already ovulated. It was the holidays, and I didn't really notice that my period was late. I was still doing my weekly blood draws from the last pregnancy loss, and no one had called to tell me my HCG levels had started to climb again. I had no idea I was pregnant for the third time. My husband and I even had a big New Year's Eve party where I had several glasses of champagne. I would have never taken a drink if I had thought there was a chance I was pregnant. I wouldn't want to do anything wrong in fear I would lose another baby.

Just a few days after New Year's, I realized I was on day 35 of my cycle. I didn't know what was going on, so I decided to take a pregnancy test. The test came up with a faint positive. I nearly dropped the pregnancy stick in the toilet. I couldn't believe it! I called the nurse right away, who told me that was not possible, that just a week ago I had a blood draw that said no HCG levels. She told me to come in right away for another blood draw. She couldn't believe it either when the test came back positive! I was pregnant for the third time.

So here we went again, an ultrasound at 7 weeks. I was nervous, yes, but a bit numb. I still couldn't believe I was pregnant again. We saw a heartbeat! I was happy, but I had been to this point two times before. I was holding my breath until the next appointment. Again we told no one that I was pregnant. Not even our parents this time. I had another ultrasound at 9 weeks. We saw the baby again with a heartbeat! Still the only people that knew of my pregnancy were my husband, the nurse, and the doctor. We just kept on praying on holding our breath. We had another ultrasound at 11 weeks. The baby had grown, and the heart was still beating away. The next ultrasound was at 18 weeks, and things began to get real. I was going to have a baby! By this time we had shared the news with our family and friends. My stomach was showing a small bump. I was so happy and so proud of my growing belly!

Ethan Alexander was born on September 3, 2004, 8 lb, 5 oz, and 21 inches long. What a long journey I went through to get him. It felt like a miracle to hold this baby in my arms. I thank God everyday for the flu, that made me ovulate late. That late egg was my Ethan. I am so lucky to have him. He is now a thriving 2-year-old, and we recently had another baby! Owen Christopher was born July 14, 2006, and is now 3 months old. I didn't have any problems with my pregnancy with Owen. I know I am so blessed to have my two boys. I look back at all the pain and sorrow that I went through, and I know that I have come a long way. I never knew that it would be so hard to have a baby. I hope that I am a better mother because of everything I went through. Somehow the terrible twos, the diapers, teething, the late nights—they don't seem so bad. My heart goes out to women who have trouble getting pregnant or who have experienced pregnancy loss, especially a molar pregnancy. It is the most difficult thing I have ever been through, and I don't wish that kind of pain to anyone. But there is hope. I have two little boys that are proof of that.

Peggy

My name is Peggy. I'm 46 years old and live in Prince George, British Columbia. I am a choriocarcinoma survivor of nearly 18 years, and this is my story.

My daughter, Carolyn, was born in June 1984, and it seemed like before I could turn around, I was pregnant with my second child, Daniel, who was born in August 1985. The two pregnancies had been very different. I sailed through my first pregnancy, but many strange things happened with the second—beginning with the pregnancy tests. They were coming back negative, then positive, then negative, then positive. Finally we received the news that the test was indeed positive and that I was 9 weeks pregnant. However, every time I went in for a prenatal appointment, the dates changed. It all seemed very confusing compared with what I had experienced the first time. I developed vertigo about midway through the pregnancy, which caused me no end of grief. There were many unusual aches and pains, but when the big day arrived, the delivery was a snap: 45 minutes start to finish. I didn't even muss my hair! My 8-lb son, however, had a variety of problems. He wasn't breathing well, his little foot was twisted, he had a scar under his tongue, and he had a wedge-shaped section of one of his grey-blue eyes that was hazel colored. These were things that could be dealt with but were just a little strange.

When Danny was exactly 2 months old, I began to hemorrhage. My husband took me to the emergency room, and my wonderful doctor suspected immediately that I had choriocarcinoma, although he did not mention that to me at the time. He tried to pass it off as a normal postpartum hemorrhage, but I knew it had been too long for that to be the case. I was bleeding so heavily that I was grossing out the emergency room nurses—how bad is that?! My doctor had seen one case of choriocarcinoma in his 30+ year career. He alerted the gynecologist, who performed a D&C later that day. Sure enough, the pathologist's report confirmed his suspicions. I was told about the disease, about how rapidly it can spread and that I would have to go to Vancouver for further tests and treatment. Vancouver is about 800 km from my home in Prince George—a 1-hour flight. I was told to expect the tests to show that the cancer had spread to my brain, my chest, and possibly my liver, but not to be alarmed because of how well this particular form responded to chemotherapy. It was too late for me to not be alarmed!

In the Cancer Control Agency in Vancouver (now known as BC Cancer Agency), I was subjected to a variety of tests. I got the bad news that my beta HCG count was above 30,000 when it should have been less than 5. I got the very welcome news that the cancer was not anywhere else in my body. I was able to begin the chemotherapy regimen immediately, within a week of the hemorrhage. Each treatment consisted of 1 day of methotrexate and 3 days of Dactinomycin, all of which had to be done on an inpatient basis in Vancouver, which meant that I had to fly there and back every second week for 3 months, leaving my husband at home to cope with the two babies (2 months and 16 months). The first dose of methotrexate I received damaged my liver, so thereafter I had to take a "rescue" drug every 6 hours for several days.

I was fortunate that the treatment was very quickly successful. The HCG numbers trended down immediately. I ended up having six bouts of chemotherapy and no surgery. I lost some hair, but because I had lots to start with, nobody really noticed. I did lose a lot of weight, and I was horribly ill during my treatments. None of the antinausea medication seemed to have any effect, but I understand that they've come a long way in this type of medication since I received it. After the final treatment I got mouth sores that nearly drove me around the bend. I think the worst side effect, though, was a sadness that stuck with me for quite awhile. I can remember looking down at the people walking by the street below my hospital room, living their lives, and feeling angry because they had no appreciation for the simple fact that they were well. I also had very little sympathy for people complaining about things that seemed so pathetic, so unimportant to me. I got past that pity party soon enough, and my disposition is better now than it ever was before. I often joke that cancer can be a good thing if you can just live through it.

It was an extremely difficult time and one that has changed me forever—for the better, I think. I don't dwell on it, but it's never far from my mind. Hence seeking out this group as a 17+ year survivor! I heard a quote once: "Cancer is but a thread in the tapestry that is my life, not the frame on which it is hung."

Rebecca

My husband, Mark, and I had been married for less than 1 year when we decided to start a family. It was October of 2001, and happily, I got pregnant right away after the first try! I took a home pregnancy test after missing my period on November 13, and shared the happy news with Mark. We were excited and surprised and wondered what the next 9 months would bring. Unfortunately, it was a rough start.

The next day, I woke up with a bad cold that developed into a sinus infection and cough. Knowing I was pregnant, I refused to take any medication to alleviate my symptoms and stayed in bed for a few days. On November 19 I was still lying around, mostly recovered from the cold but still coughing. I remember we were watching television in bed. I had a pretty violent coughing fit, and all of the sudden I felt this shooting pain in my lower right abdomen. I figured it was a small muscle pull from coughing so hard while lying down. Nevertheless, the pain went away, but about an hour later when I got up to go the bathroom, I was spotting. It wasn't very much—just a very dark reddish-brown stain— but it was both heartbreaking and terrifying at the same time. For the 6 days since I'd gotten a positive on my home pregnancy test, I was so thrilled to be pregnant. I had already begun planning and had already shared the good news with my parents. Seeing that little bit of blood on my underwear brought my exhilaration to a crashing halt.

I didn't know what to think. I was convinced that my coughing had caused the shooting pain and the bleeding, but

I didn't think coughing could make you lose a pregnancy, if that is what was happening. The next morning, I called the obstetrician first thing. The on-call nurse advised me to watch the bleeding for a few days. She said that as long as it wasn't bright red and didn't contain clots, and as long as there was no pain associated with the bleeding, then it may resolve itself and the pregnancy may be okay.

I continued to spot for the better part of a week. It wasn't much, but it was very persistent, and it was still just a brownish red. So on November 26, I called the doctor back. The doctor asked if I could come in right away for a transvaginal sonogram to see what was going on. That afternoon, I met with the on-call doctor, who performed the sonogram. She didn't like what she saw and told me as much. She told me she couldn't see a defined gestational sac and couldn't really discern a fetal pole, and she thought that she should've been able to see more for me being almost 6 weeks along. She scheduled a more detailed transvaginal sonogram at the lab for 2 days later, and in the meantime she ordered a beta HCG blood test.

Over the next 2 days, the bad news came fast and furious. My first blood test came back with my HCG measuring at 52,800. My doctor felt that that was very high and advised me to start thinking about a D&C. I thought she was crazy; after all, those numbers seemed normal to me after researching it a little on the Internet. I didn't know what to think of the sonogram — maybe it was just too early to see anything. But the next day, at my second sonogram, my worst fears were confirmed: there was no embryo. My second blood test revealed that my HCG had risen to over 60,000. My doctor called me again to explain my options, and this time she didn't just advise the D&C, she ordered it. That day was the first time I heard the words "possible molar pregnancy." We scheduled my D&C for December 11, about 2 weeks away.

During those 2 weeks I continued to bleed, but I also continued to feel very pregnant. I was nauseated and fatigued, and I couldn't believe the pregnancy wasn't viable. I resigned myself to trust my doctors. On the day of the D&C, my doctor told me prior to the surgery that she was going to examine the products of conception herself and then send them to pathology for further testing to rule out a hydatidiform mole. So on that day, at about 8 weeks along, I was wheeled into an operating room, and my real ordeal began.

The surgery itself and the recovery were easy—I had no heavy bleeding and no real pain. Mark and I went to my parents for the Christmas holidays to relax, regroup, and recover, both physically and emotionally. Two days after Christmas, I had a follow-up appointment; it had been 2 weeks since the D&C. I was still bleeding a bit, but I expected that after the surgery, so I wasn't worried. So I was very surprised at what my doctor had to tell me. She didn't mince words—pathology had found evidence of molar tissue and no fetus, so I was diagnosed as having a complete molar pregnancy. I was shocked and scared. I barely heard her explain the next few weeks: weekly blood draws until my HCG levels fell back to zero, followed by at least a 6-month wait period before we could try to conceive again. I got dressed slowly, my head spinning, and went straight to the lab for my first postoperative blood draw, which marked my levels at 1,987, a huge drop from the 60,000+ before the D&C. My doctor was confident it would drop quickly.

The next few weeks were a blur. Amidst the waning Christmas holidays and the New Year's Eve celebrations, I was up to my neck in literature, books, research—anything I could get my hands on about molar pregnancies. I didn't find much. Hardly anyone I spoke to knew anything about them, and there was precious little information on the

Internet and at my local library. I also was still bleeding continuously. It wasn't heavy, but it was very persistent and very annoying. I shouldn't have still been bleeding from the D&C. In a way, I knew things weren't good.

I went for my second postoperative blood draw on January 4, 2002. The results weren't encouraging; my levels were now at 2,105. My doctors were cautious at this rise, but let it go for another week. On January 11, my levels had risen to 3,773. I was called in immediately for a sonogram to check and make sure I wasn't pregnant. I wasn't. Two days later, they checked my levels again, and they were up slightly to 3,778. I remember this day like it was yesterday. I was sitting at work, and it was late in the afternoon. It was a cold but clear January day, and the sun was setting. My husband worked at the same office, and we were getting ready to go home; I was staring out the window waiting for him so we could walk out to the car together. But then my phone rang. It was my doctors' office, a receptionist that I knew well by now—a woman with friendly compassion who'd been there through all of this. She had a lot of instructions for me. She told me to go immediately to the pharmacy to pick up prescriptions in preparation for an early morning CT scan of my brain, liver, uterus, and lungs, and then go to the radiology unit at the hospital to pick up the contrast fluid I would have to drink first thing in the morning. My appointments were already set up; she'd taken care of that. I could hear the urgency in her voice. I was petrified. My husband appeared at my office a few minutes later and found me pale and crying and thinking I had cancer. We walked out together, finished the business at the pharmacy, and went home to prepare ourselves for the days to come.

The next day, thanks to the urgency of my situation, I had the results of my CT scan only an hour after I returned home from the scan. My parents had driven down to our house to sit and wait with us. The phone call was good news: there was no metastasis. But I would still have to start chemotherapy the next day to combat the growth that apparently was confined to my uterus.

For the rest of that day my parents stayed with us, and we coped as well as we could. There were a lot of unanswered questions about my prognosis, how long the treatments would last, and what the side effects would be. But the biggest unanswered question was the future and my chances of having children. We were all sad, apprehensive, and shell shocked, to say the least. We had gone from being so overjoyed at the prospect of a child and grandchild entering the family to having to face what nobody wants to face: chemotherapy treatments and a terror of the future.

The next day, I received my first chemo shots. It was methotrexate, and the doctor told me it usually didn't have bad side effects. The worst, she said, would be dry skin and lips. I was to have weekly blood draws and chemo shots until my levels fell back to zero. The shots weren't bad—they were administered in my backside, one on each side each week. Although I did have very dry skin and lips, it wasn't all that bad. The worst side effect was something I was completely unprepared for: I was overwhelmingly tired.

After my first round of shots, my numbers came down from 3,778 to 3,409. My doctors were very encouraged and urged me to get lots of rest and be positive. I was feeling okay, despite the fatigue, so after my second round of shots my husband and I decided to take a trip to Florida. We lived in Maryland at the time, and it was the middle of winter. We had been invited to attend a relative's wedding in Florida and had originally planned on going, but we had all but cancelled our reservations because of my condition. At the last minute we decided we could use the break and a change of scenery (and climate!), so we packed our bags. The only problem was finding someone who could administer my chemotherapy shots while traveling. I called every hospital and clinic in the area of Florida to which we were traveling, but no one would agree to give me the injections, even with my records and a note from my doctors. We almost had to cancel our trip, but my mother-in-law, who is a registered nurse and who was also attending the wedding, finally agreed to give me the shots. The day before we left I got another round of shots and would have to call from Florida to check the results before going ahead with the subsequent round, which I carefully packed in my suitcase, along with two brand new hypodermic syringes.

Mark and I hopped in our car and spent the next 2 days driving to Florida. It was wonderful! We talked along the way, played music, stopped when we needed to and sometimes when we didn't need to. The weather grew warmer and sunnier as we drove south. When we arrived in Florida, our first stop was Jacksonville, where we decided to get some lunch and go for a walk on the beach. Incredibly, it was on this beach that I decided to call my doctor and find out the results of my previous blood test—the one I had drawn just prior to getting the round of chemotherapy the day before we left for Florida. I can still remember the day—it was warm, sunny, and breezy, the air was fresh and bright. I was sitting in the car with the door open, on the cellphone, watching the waves roll in as the nurse told me my levels had fallen all the way down to 363! I hung up and dashed through the sand to tell Mark. Although we weren't out of the woods yet, we were on a nice 10-day vacation, and it was the best news we'd heard in months!

From that day onward, things quickly improved. My mother-in-law injected me again the next day with no problem, we attended the wedding and had a wonderful time, and we then spent the rest of our vacation enjoying the weather and relaxing. In Clearwater, Florida, 3 days before we left, I finally stopped bleeding. I had been bleeding since November 19, and it was now February 7. Somehow, I knew that we were near the end of this ordeal.

With light hearts, we drove home, and even the return to dismal and cold weather couldn't break our spirits. We went straight to the doctor for an immediate blood draw and waited the hour or so for the results. When they came, we were thrilled: my levels had come down to 21! Although I still needed another injection, my doctors felt it would be the last one, and they were right. The following week my levels were down to 4, and I was pronounced healthy. They instructed me to come in for monthly blood draws for a year, after which I would be cleared to try to conceive again.

I suppose in a way, it was a bit daunting. After months of living on eggshells, of constant phone calls to and from the doctor and to and from family, waiting, always waiting for the results of one test or another, we were free. We found ourselves in the position of not quite knowing what to do with ourselves. So we decided to do some traveling. We realized that once we had a baby, we might not be able to go anywhere anytime soon, so off we went.

In early May, we took a 10-day trip to San Antonio and South Padre Island, Texas. In September, we visited family in Jackson, Wyoming, and went to Yellowstone National Park for a few more days. In the spring we were planning a long, relaxing trip to Alaska. Needless to say, the months went by quickly. We were working long days and going on lots of trips—besides the long ones, we also did weekend trips to visit friends and family, and we sometimes did some day trips around Maryland. Before I knew it, Christmas had come again, and New Year's, and in late February my doctor called with my last blood test result and told me we were free to try to conceive again.

Because we already had plans for our trip to Alaska in early June, we put off trying to conceive another few months, something I wasn't happy with initially. But I didn't want to go to Alaska pregnant, so we decided that a few more months wasn't that long in the grand scheme of things.

We packed our bags again and embarked on a 17-day trip to Alaska, which included an 8-day cruise from southeastern Alaska to Vancouver. As eager as we were to go on this trip, we were even more eager to start making plans to make a baby again as soon as we got home. Alaska was beautiful, more beautiful than we ever expected. And inspiring, I suppose…although we weren't trying to make a baby yet, we certainly weren't preventing it, and several weeks after we returned home I missed my period. To our shock and delight, we discovered I was pregnant.

I was flooded with mixed emotions: doubt, joy, fear, wonderment, terror. I barely allowed myself to think about a little baby, but I couldn't help but wonder if the tiniest little glimmer of hope I had wouldn't let me down. About a week later, my world crashed down again—I started spotting. I was so devastated. I called my doctor in tears, but there was nothing they could do to help alleviate my fears—it was too early. They scheduled me for a sonogram at 6 weeks, 3 days, which was more than a week away! Nobody, not even my husband, my mother, or my best friend knew what I went through those next days. Nobody knows how many tears I cried, mostly in the shower, the only place I knew I'd be alone and undisturbed. Nobody knew the anguish I faced as I prayed to, bargained with, and challenged God. And nobody could possibly know how close I came to losing my faith.

When I arrived at the sonogram a week and a half later, I was a wreck, mostly emotionally, but physically too. I hadn't slept much, I was still bleeding, and I was dead tired. I was nervous to the point of passing out, my palms were sweaty, and I could feel my heart pounding in my chest. But the news was good—there was a little embryo, with a strong little heart, beating away. I burst into tears and poured out my story to the radiologist, who listened with love and gave me a hug.

Although I faced other scares during my pregnancy, the bleeding finally resolved, and I went on to carry a healthy son to term. I delivered him on March 10, 2004. He is the light of my life, my little miracle, God's perfect blessing. Since then I gave birth to a second son, on June 22, 2006, and I enjoyed an uneventful and easy pregnancy and added to the already incredible blessings in my life.

My boys are my inspiration, my reason for life. My husband and I are devoted to them in a way I never knew parents could be devoted. Mark and I share a closeness borne of fear of tragedy, and now share the miracles and blessings of having not only survived, but survived to tell the story and hopefully help and be an inspiration to others. Sometimes I wonder if fate doesn't know better than we do; we grew so much during our ordeal, it's made us who we are and determined what kind of parents we are. It may have been a hard lesson to learn, but looking back, I can truly say it's made me a better person.

Sharon

In May 2000, at the age of 36, I discovered I was pregnant. I wasn't too sure how I felt about being pregnant at first but grew to love the idea within a week. I'd had an idea for a couple of days that I may have been pregnant, because I'd been a bit nauseated in the mornings and my breasts were ultra tender to the touch. I remember saying to my best friend that something didn't feel right and that I was terrified of losing the baby. She asked me what I meant when I said something didn't feel right, but I just couldn't explain to her what I meant. I seemed to have a lot of morning sickness, which is unusual for me, and my stomach seemed to grow so huge so quickly. None of this seemed right, because I hadn't experienced these symptoms in my earlier pregnancies.

I was 11 weeks pregnant when I started to bleed. I rang the hospital and was told to rest with my feet up because the bleeding usually stopped and because there was nothing that could be done anyway. The next day I was still bleeding, but it had gotten heavier, so I went to the hospital. They checked me out and said I should be fine, not to worry, go home and rest. The day after that I'd had enough and went to my local doctor. He gave me a referral to see an obstetrician, who in turn sent me for an ultrasound. That afternoon I was booked into the hospital for a D&C the next morning. I had no real idea what was going on, only that my baby didn't have a heartbeat and that a bunch of "grape-like things" had been found. I was absolutely devastated. Funny how you don't realize how much you want something until it's taken away from you.

On the morning of July 19, 2000, I had my D&C. I was told afterward that I had had a partial molar pregnancy and would be required to have blood tests weekly until my HCG levels had returned to "normal." Luckily for me this happened within about 7 weeks. My doctor said I was free to try again anytime I wanted. I still had no real idea of what had happened to me so I turned to the Internet, where I found the information that my doctors wouldn't, or maybe couldn't, tell me. I decided to wait for the 12 months that American women are advised to wait before trying to conceive. I needed time to consider the risks of having a child at my age, and I wanted to make sure I was having another child for all the right reasons, not just because I wanted a replacement for the one I had lost.

In August 2001, at age 37, I discovered I was pregnant again. I was cautiously optimistic and quietly ecstatic. I had ultrasounds and bloodwork done right away and everything looked normal. I have four other children, a daughter born in 1982 and three sons born in 1984, 1986, and 1988. The child born in 1988 was one of twins, but I miscarried one of the twins about 8 weeks into the pregnancy. In 1983 I had a child diagnosed with anencephaly, which is a condition in which the brain doesn't form or grow properly. This was diagnosed through routine ultrasound at 16 weeks. I had to have the pregnancy terminated, but by the time everything had been arranged I was 20 weeks along and had to actually go through labor and give birth naturally. The child was a boy. In 1994 I miscarried at 9 weeks, but there was no apparent reason for the miscarriage.

At 37 weeks into the latest pregnancy I went to my obstetrician for a checkup. The baby had been head down for quite a while but was now laying transverse. At 38 weeks she was head down, at 39 weeks she was breech, at 40 weeks transverse, and at 41 weeks head down. We decided to induce while she was in the right position. On Thursday,

April 25, 2002, at 8 A.M., I had my water broken. Mild contractions followed immediately, but things didn't start getting serious until about lunchtime. I hopped into the bath at 2 P.M., and this eased the pressure of the contractions. My sister was beside me for the whole labor, and she was great. At 5:15 P.M. I told the midwife I wanted to push and asked if she wanted me to get out of the bath, but she told me I could stay where I was. At 5:55 P.M. Arizona Faith entered the world. She was 7 lb, 5 oz, and 20 inches long. She was born underwater, because the midwife said she was happy to do the delivery in the bath as long as I was comfortable with it.

When I had my molar pregnancy I was devastated, as I'm sure all women are. With all the bloodwork and everything else that goes along with it, you begin to wonder if you'll ever have a baby. I wrote this because I want you all to know that there is life after molar pregnancy.

Susan

To make a long story short, here is a little about me. After almost a year of fertility treatment I finally got pregnant in March 2001. During my first sonogram, I was told that I was carrying twins, which is what my husband and I were hoping for. Because of the fertility treatment and my size (5 ft tall, 120 lb), they watched me very closely. I had sonogram after sonogram only to find out that one twin had stopped developing and that I might start bleeding because my body would have to dispose of the placenta naturally.

Although upset that I lost one, I was still very grateful to be pregnant with the other! At 6 weeks' gestation I started bleeding. I was not too worried because I expected that. On June 6, 2001, I went to the doctor for an examination and was told that the baby looked great, you could see arms and legs, eyes and nose, everything was perfect except my uterus was very large. After a long exam, I was sent straight over to a specialist for another opinion of the enlarged uterus. That's when we got the news that I will never forget. The placenta of the nonviable twin looked like a cluster of grapes and was very large. I was diagnosed with a molar pregnancy and had to have surgery immediately to remove the tumor.

Unfortunately, there was no way to remove the tumor and leave the healthy twin in place, and I couldn't wait a few more months until the baby was large enough to live outside of me either. I had no choice but to terminate the pregnancy. The day of surgery my HCG level was 1.6 MILLION, and my thyroxin level was 25. I was a very high risk patient at that time. During surgery, I lost 500 cc of blood, and my tumor was the size of a cantaloupe. After a day and half in the hospital, I was released to go home to begin the waiting game. My levels were dropping slowly but surely, although they had a long way go. I tried to keep a very positive attitude until July, when my levels were on the rise. I went straight to a reproductive oncologist and started chemotherapy that very day. I had 35 treatments of chemo—5 days a week for 7 weeks. I alternated each week between two medications and finished chemo on September 29, 2001. One of the best days of my life!

It was at that time the waiting game began all over again. As of September of 2002 I was clean and clear, ready to get pregnant again. On February 12, 2004, I am very grateful and blessed to say that our family grew by three! I delivered healthy triplets (two boys, one girl) on October 30, 2003. At birth, Blake and Grant each weighed 3 lb, 15 oz, and Chloe weighed 2 lb, 13 oz. Grant is now 13 lb, 8 oz; Blake is 12 lb, 5 oz; and Chloe is 9 lb, 6 oz. Although my pregnancy with the triplets was extremely successful, it was not without fear. I had my first sonogram when I was 6 weeks pregnant, and I had three sacs and three placentas but only two heartbeats. This is what I feared the most when I found out I was pregnant again. However, I thank the Lord above for the miracles He gives us, because the next week the third heartbeat was there. I carried the triplets to 31 weeks and 3 days and had the most wonderful experience without complications. I would do it all over again in a minute.

For those of you who are experiencing the pain and loss of a molar pregnancy, please be positive, have faith, and don't give up. I was very nervous about getting pregnant again after the mole, but my love for children outweighed the potential for another molar pregnancy.

Tamika

In January of 2006 I became pregnant for the second time. My 2-year-old was dying to be a big sister, and we were all so excited. I called to set up my first appointment. About a week prior to the appointment (6 weeks pregnant) I started to spot. They had me come in for an ultrasound. After a long silence, a doctor came in to explain that I had what appeared to be implantation bleeding. They told me to avoid picking anything up, stay off of my feet, and take progesterone. I had another appointment a week later.

At the next appointment they said the bleeding had stopped and the pregnancy appeared to be saved. They saw a fetal pole both times, and it had grown, but they still wanted me to come back in a week. At the third ultrasound I was told that the baby had died, that I should stop taking the progesterone and either schedule a D& C or let nature take its course. I was past 9 weeks, and there was no heartbeat. I had read stories of people being told they would miscarry and that not happening, so I decided to let nature take its course.

I waited for 7 weeks and became very sick. I decided at that point to schedule a D&C. This is when I was diagnosed with a missed miscarriage. I had the D&C on March 31. I felt better the very next day. I had 1 good week, and then I started to experience severe pain and to pass blood clots ranging in size from a quarter to a fist. I had to spend a night in the emergency room as they monitored and collected the clots. I went to my postoperative appointment the next day. My doctor took some blood again that day to see if there was an explanation for the clotting. She also told me at that appointment that they tested the material removed in the D&C, because it was more than usual. She told me the tests had revealed that I had had a molar pregnancy. She wanted to monitor me because the cells often regenerate.

The next day my doctor called me at home to tell me that the results of the bloodwork revealed that I needed to see a specialist. I had read through all of the information she had given me the day before. I was aware there was some risk of cancer. She told me my beta HCG had more than doubled since the day of the D&C, and as a result she had scheduled me an appointment with the best specialist she knew, the head of OB/GYN Oncology at the University of Oklahoma. She tried to comfort me by telling me that this is who she would send her sister to if she were in my situation. My appointment was that Thursday. I spent the next day online and in tears. I was angry. I had index cards full of questions needing explanation. I talked to my doctor on the phone several times and at least once for over an hour. She was very patient and did all she could to answer my questions. I wanted to know how they missed that I had this rare pregnancy when they were looking at it weekly? Why didn't I miscarry? Had I put myself in greater risk for cancer by waiting 7 weeks to miscarry?

I will never forget the first time I walked into the Virginia Cade Cancer center at Oklahoma University. I was so overwhelmed, and I had already realized that they were going to talk to me about cancer. Then I had 13 pages on a clipboard to fill out all about the pregnancy and releases for ongoing cancer research, etc. I had not even made a dent in all of that when they called me back. They took me to a very comfortable room with couches and magazines, offered me juice, and said the doctor would be right in. I was impressed at how comfortable they tried to make me feel. A fellow came in and answered questions and told me what they knew about me and explained what would happen next in the appointment as well as the likely course of action in my case. She told me that I had had a partial molar pregnancy, that it was very hard to detect and had been discovered at the earliest point they had seen. They asked who my doctor was and told me that she had gone above and beyond what was required and that is why I had a good chance of quick treatment as well. This helped calm the anger that had been burning in me about being in this situation. She explained the partial molar further, telling me that in this case two sperm fertilize one egg at the same time. Instead of twins, there are too many chromosomes. The body continues to try to make a baby, but the recipe is off. So there were what she called "fetal parts," but they would never really make up a child. She said there had been cases where under these circumstances there have been births, but none have survived. So now I had a regeneration of those cells. I would most likely need to undergo chemotherapy.

The doctor came in and answered more questions, explained more fully, and then we all went to an exam room. He examined me, and then we went back to the meeting room. He said we may never know if it had already become cancer; they took some blood tests and said that cancer or not, I would need to start chemotherapy. I felt okay about the appointment and went home hoping that the numbers had not gone up. I was home an hour when they called and said I had had a 75% rise in my beta HCG number and needed to come back that same afternoon or early the next morning to begin chemo. I asked, is it cancer? The nurse who called said she didn't know, but I needed chemo regardless. So the next day I went in and met with another doctor, because mine was gone to speak at a conference. This other doctor walked in and said she had good news and bad news, but that it would all be okay. She told me that if you have to have cancer, this is the one to have, because it was the first kind cured, and no one dies from it anymore. Then she explained the process of chemotherapy and asked me to participate in a clinical trial studying this form of cancer. The trial consisted of two drugs that were randomly assigned to kill the cancer. The risks for one were too much for me to handle, so my husband and I decided to choose the drug methotrexate and take the safer route. I was going to get two shots weekly, one on each side of my buttocks.

I walked into the infusion room, and I remember the neon yellow bottles filling up and then a little pinch on each side. When I walked out of the room, I felt no different, but when I went to the elevator I burst into tears. My husband asked what was wrong, and I said I realized my life will never be the same. I took methotrexate for more than 14 weeks. My numbers were impressive at first, but then plateaued, and then started to rise. We needed a new plan. They offered me a different clinical trial this time, because this was a resistant case. They said I would be done with chemo in a month or two, so I started on a drug called Alimta. This was administered by infusion in 3-week cycles. I was so sick with this drug, my skin darkened, I lost 14 pounds in about 8 days, and I had sores all over my mouth. I couldn't eat or stay awake, developed a bright red rash, swelled up, and had a fever for 7 straight days. The drug was working, but I was responding to it in a negative way unlike anything they had ever seen.

I then developed a very bad cyst and ended up in the hospital again. I survived the days by taking about six supplemental medications to manage the symptoms. I had three cycles of this drug before they declared it a failure too.

My doctor was consulting with other doctors around the nation, and I was regularly being seen by several in the field. I had to undergo a couple weeks of tests; they checked my brain and lungs to see if it had spread, but I was okay. They decided to try dactinomycin, the other drug in the very first clinical trial that I had passed up. I started this drug, also intravenously, but in 2-week cycles. I was counseled at this point to prepare for the hair loss and was told to get a wig within the next few days. I went with my mom and 3-year-old daughter to buy it. This was a very emotional experience. I had thinning of my hair about 50%, and most of fit at the top and around my face. I never needed my wig, and finally the chemo brought my levels to zero. I had my last treatment of chemo in December. I am halfway through my 12-month wait. My doctor said this was one of the most complicated and resistant tumors they had ever seen. I was told at my last appointment that I should be able to conceive again in January of 2008.

Vicki

My story begins many years ago. My husband and I met while still in middle school. He was a year younger than me, and other than the fact that he was fun to hang around with and easy to talk to, I didn't think much about our friendship. It wasn't until we were in high school that things got serious. We dated for most of my sophomore and his freshman year. It was such a serious relationship, and I had such high hopes for my future, that I was scared witless by it. So I broke up with him, but we stayed friends, and he always had a special place in my heart. That didn't change even when I married someone else, and he joined the Navy. We each lived our separate lives, but we remained in contact, and my feelings for him never really changed. Six years later I was living on my own with three young children after separating from my first husband. I was struggling to raise my kids, attending college full time, and working part time. When I sent a Christmas card to his parents' house, I truly never expected to even get a response, so I was shocked when he called one night and said he was in the area and wanted to stop by. That first visit led to a weeklong vacation, much of it spent with my family. It wasn't long before he decided to move back to where we had grown up and where I still lived.

Those early years of our relationship were very rocky. One of the reasons I had broken up with him in high school was because I knew even then that he didn't want children. I had always known I wanted to be a mother, and that I wanted to have my children while I was young, so this was a huge obstacle to me. For him, getting involved with me when I had three kids under the age of 6 was very difficult. He didn't know how to be a parent and had no desire to learn. It was tough for me as well. I told him on more than one occasion that we were a package deal, and if he wanted to be involved with me he had to take the kids too.

Everything changed about 2 years into our "new" relationship. Things had not been going very well when I found out I was pregnant. We were both very disturbed by the news. I was just beginning my junior year of college and this would be baby number four for me. As a single parent with limited income, I didn't know how I could do it. Leo told me he didn't think he was ready to become a parent. After much soul searching I knew that no matter what, this baby was precious. I told him that he could choose whether to be involved or not but that I already knew how to be a single parent and would figure out how to do it with four kids instead of three. It took a while, but his becoming a daddy to our son was the best thing that could have happened to us.

I won't say that everything was easy once Anthony was born, but we managed. We moved into my mother's house when Anthony was a little over a year old and rented it from her for 3 years before I was finally able to buy it from her. The following year, we had a beautiful wedding in our backyard, with my daughter as the maid of honor, my older boys as junior groomsmen, and our 5-year-old son as the ring bearer. It was just a couple of months after the wedding that the idea of having another baby first came up.

We were attending a Christmas party at a friend's house when someone arrived with their new baby daughter. I truly thought we were done having babies and never expected to hear my husband mention adopting a little girl. His original idea was to wait until our daughter left for college. As strange as it sounded at the time, his offhand comment led to many discussions and ultimately a decision to try to have another baby. We decided that even though we had three boys and really wanted a little girl, we would try to have a baby together before pursuing adoption. To this day, I am in shock that it was my husband who pushed for another baby.

With all my older children, I never really had to try to become pregnant, so all of the trying to conceive stuff was brand new to me. I went online and started reading everything I could about it. I stumbled across a message board for women who were trying to conceive, pregnant, or mommies. Little did I know that board would become my support line in just a few short months. The first 3 months of trying didn't produce a positive pregnancy test, so I decided to begin charting my cycles. Wouldn't you know that the very first month of charting I got my positive!

We were all so excited to have a baby on the way. Even the older kids were ecstatic. My teenage daughter, who I was worried would be embarrassed, thought it was great. The two boys in the middle said they didn't really care, but they were just as excited as we were. The little one was the cutest of all. At 6 years old he couldn't wait to not be the baby anymore. My first visit to the obstetrician at 10 weeks was uneventful. Everything seemed to be going according to plan. And why shouldn't it? After all, I had already had four healthy, uncomplicated pregnancies. My biggest worry at that point was that this new baby would be large, because Anthony had been 9 lb, 13 oz at birth.

The first clue that anything was wrong came at my routine 12-week ultrasound appointment. The tech wouldn't let us see the screen and kept asking if there was any way my dates could be off. Of course because I had been charting I knew exactly when I had ovulated. If anything I should have been a week farther along because I had actually ovulated a week early that month. Finally, the technician said she wasn't seeing a yolk sac and let us see the screen. All I remember is seeing lots of black space and an empty oblong amniotic sac. We were asked to sit back in the waiting room until the nurse practitioner could speak with us. I knew something wasn't right, but I had no idea my world was about to fall apart. The nurse practitioner took us to the doctor's office and said she thought it was a molar pregnancy, but even if it was something else, it was clearly not a viable pregnancy. She sent me immediately for bloodwork and said she would call in about an hour with the results.

My husband had no idea what a molar pregnancy was, and all I knew was that it was mentioned in my pregnancy book. I clearly remember telling him on the way home that the good news was we would get special treatment the next time I was pregnant. I was in shock at that point and the reality of the situation hadn't yet sunk in. The hardest part of that day wasn't the horrible blood draw I had to endure but telling my kids that the baby had stopped growing, and we wouldn't be having a new brother or sister after all. We had planned on picking them up after our ultrasound, so it was all still a blur at that point.

We got home and I immediately went to my book and started reading about molar pregnancies. I was shocked by what I read but still trying to be optimistic. I was scheduled to work that afternoon as a tutor, but the phone call from the doctor's office changed all that. The nurse practitioner called back and told me not to eat anything because my HCG was extremely elevated, although she wouldn't give me the number, and I was scheduled for surgery that afternoon with the doctor on call because my doctor was out of town that day.

I called my job and told them I would be back on my next scheduled day, 2 days later, and then I had to call my mom. Little did I know I would never return to that or the other job I had that summer.

I think if the doctor had been able to do my D&C that day, things would have been a little easier for me because I hadn't yet had time to think about any of it. As it turned out, they cancelled my surgery because the doctor was too busy to do it. That gave me time to research both the procedure and molar pregnancies. My own doctor never called to check on me that day, so I called the doctor on call with my concerns. One of the biggest problems I had was a horrible phobia about general anesthesia. He assured me that the D&C could be done under a local or regional anesthesia and confirmed that this was indeed a molar pregnancy. I wasn't so convinced. I still have my doubts, although I know without a doubt it wasn't a viable pregnancy.

Because this doctor wasn't able to perform my surgery, I was back to using my obstetrician for the D&C. Problem was, she refused to even consider doing it without general anesthesia. So did every other doctor in the county. So I refused to have the surgery done. I was angry, hurt, scared, and very difficult. My best friend, who had just a few weeks earlier discovered she was pregnant a month after her husband's vasectomy, came to my house to talk me into going to the hospital. I wouldn't budge. My baby was dead, nothing else mattered. I didn't care what happened to me.

I am not proud to admit it, but not even my other kids could snap me out of my funk. My 10-year-old decided he wanted to quit football and I didn't even have the energy to talk him into sticking it out. He went back to the team a week later, but we were still dealing with the backlash from that more than a year later. My relationship with my husband took a nosedive. He tried to bully me into doing whatever the doctors said and wasn't even listening to what I was saying. I finally agreed to let the first on-call doctor do the surgery, but he was on vacation now for 2 weeks. I was willing to wait. Everyone worried that I would hemorrhage while waiting, but I hadn't even had any spotting, so it wasn't a concern to me. Besides, I really didn't care at that point. I was holding on to the only scrap of control I had left.

It still bothers me that no one listened to why I was so dead set against being put to sleep for the surgery. I had been molested by a cousin as a young child and since then have had panic attacks. The idea of not knowing what is happening to me sets the attacks off in a major way. I had talked myself into having general anesthesia in my early twenties when I needed my wisdom teeth pulled, and that experience was enough to convince me I would never do it again. Everyone tried to tell me anesthesia was better now, and I wouldn't have the side effects I had before. No one understood that it wasn't the side effects I couldn't deal with, it was the not knowing. I couldn't face fading out and then waking up hours later with no knowledge of what had happened. Those 2 weeks of waiting were horrible. I did nothing but surf the Web, looking for any information I could find about molar pregnancies and their treatment. None of it was good. My doctor wouldn't listen to anything I found anyway. I was still angry at the world and everyone around me. My husband and I fought daily. He didn't understand what I was going through, and I didn't feel he was being supportive. He just wanted me to get it over with and move on. It wasn't that simple to me. To make things worse, every time I had to have blood drawn was worse than the time before. My veins had never been very good, but now they would collapse as soon as the needle entered my skin. Just getting enough blood for all their pre-surgical tests was a nightmare.

Then came surgery day. I had already had to reschedule it once so that I could go to two job interviews. There was no way I could face going back to my jobs as a summer tutor or working with preschoolers, and our finances were going downhill fast. I had already been off work for almost 3 weeks, and I knew I had to get a new job. So I rescheduled the surgery for 2 days later. I didn't want to go, and I was further irritated by the fact that they expected me to go to the doctor's office for a preoperative appointment at 8:00 A.M. but not be to the hospital until 1:00 P.M. What in the world did they think I was supposed to do in between? I still don't understand why they couldn't do the preoperative stuff right before the surgery, and I don't think I ever will.

The morning of the surgery I was not easy to get along with. I had told my husband the week before that if I told him I didn't want him there it just meant that I was scared, and he needed to be strong for me. He didn't remember that, so when I lashed out at him, he wasn't there for me. I drove myself to the doctor's office and forced myself to endure that visit. When they asked if I was waiting for my husband I told them no, and they had better not allow him at the hospital either. After the visit, I sat in their waiting room and read a book for 3 hours. I knew that if I went home I would never go to the hospital for the surgery. I left just before they closed the office for lunch, but I still couldn't go to the hospital for another hour. I went and bought another book to take with me and couldn't stand not drinking anymore. I bought a bottle of water and took a few sips. I hadn't had anything to eat or drink in almost 24 hours at this point.

By now I was really angry at my husband. The whole time I was at the doctor's office, I kept expecting him to show up but he never did. I called him from my cellphone and told him I hoped I didn't make it out of surgery and he wasn't there. Since I had nowhere else to go, I went up to the hospital and sat in the waiting room. No one asked me why I was there or even noticed that I had come in. I didn't see anywhere to sign in, so I just kept waiting for them to notice me. They never did, but the doctor's office did call and tell them I was being difficult. I got even angrier when the nurse at the front desk told another nurse that she could make me do whatever she wanted me to. She obviously didn't have a clue what she was dealing with. I must have sat there for almost 2 hours before I went to the desk and signed in. I was being difficult and didn't trust myself not to start screaming, so I was choosing not to speak at all. The nurses didn't like that. The first nurse to take me back to prepare me for surgery tried to force me to talk, and I walked out of the pre-op area. I didn't know what to do next so I just sat back down, and another nurse came to talk to me. I ended up telling her, "I don't have to be happy to be here, and I don't have to talk to you. I am here, and that is the best you are going to get." Needless to say, they found a new nurse to prep me. Not that it did any good. I told them that my veins were horrible, and they didn't believe me. While trying to numb my hand, she hit a nerve and my hand jumped, so the nurse decided to just put the line in without numbing it first. She sat there and told me how great my veins were while I screamed in pain and my veins collapsed.

The anesthesiologist came to see me because I had refused to sign consent until I had spoken with her. I wanted to make sure they knew I did not want general anesthesia. She ended up having to help them hold me down for almost 10 minutes while I screamed bloody murder as they tried to get an IV into my arm. As soon as they stopped poking me I started taking deep breaths and trying to calm down because I knew if my blood pressure was too high they would sedate me. I even told the anesthesiologist I couldn't understand how I could give birth four times with no pain relief but this hurt so much. As soon as the IV was in, they had me sign consent and took me to the operating room.

When we entered the operating room, the nurse told the doctor that they had given me Versed. This made me mad because I had specifically told them I didn't want to be sedated and because they had given it to me without asking or telling me about it. There wasn't anything I could do about it at that point. I am still angry that they gave it to me, because it made it very difficult for me to focus. As a result, I had a very choppy memory of the surgery, which was what I was trying to avoid in the first place. Many times after the surgery I woke up in the middle of the night in a panic because I was trying to piece together what happened. These memories were like the ones I have of being molested as a child, and they all mesh together. It was not a fun experience.

Because I had driven myself to the hospital and had had a spinal block, after the surgery I was taken upstairs to a room. They asked me if my husband could come visit; apparently he had gotten to the hospital sometime after I had been taken into surgery. Due to the sedative I told them I didn't care. He only stayed a few minutes and then left to take the kids to football practice. He came back later only long enough to pull his car around and load me into it. He then drove me across the parking lot to my car, and I drove myself home. I wish I could say that after the surgery everything was fine, but it wasn't. I was angry with him for a long time, and in some ways, I still am. He wasn't there for me when I needed him most. Nobody came to visit me at the hospital because he told them I didn't want anybody there. I felt very alone. Especially after the surgery, no one wanted to talk about what had happened. It was as though it never happened to everyone except me.

The hardest part was the fact that my doctor kept telling me I had to wait a year to even think about getting pregnant again. My daughter was already almost 14 and my youngest son had just turned 7. I didn't want to wait a year to try again. Some days I didn't know if I even wanted to try again and other days I wanted it so bad I couldn't see straight. The nursery that I had started painting and collecting baby furniture in was right next to my bedroom. It didn't have a door, so I stapled a sheet over it and wouldn't let anyone go into it. Just seeing that sheet was enough to rip my heart out. For a baby I didn't know I even wanted, it sure tore me up to know it was gone. I just didn't understand how this could happen.

There are two things that helped save my sanity. One was finding the molar pregnancy support group online. It helped so much to have people to talk to who knew exactly what I was going through and to answer my questions. Everyone in my life expected me to be sad but couldn't deal with my anger. Only my online friends knew how I felt. I still visited my pregnancy message board and got a lot of support there, but I couldn't deal with all the pregnancy and trying to conceive talk so I didn't go there as much. The other thing was getting offered a new job. I was supposed to hear back from one of the jobs I had interviewed for on the day of my surgery. Because I didn't hear from them I assumed I hadn't gotten the job, and I was really bummed about that. I had been trying to get hired at a school for almost 4 years, and I really wanted that job. When they called me to offer me the job 3 days after my surgery is when I started to heal. It was a turning point for me, and I knew I had to do it. I had to go into the principal's office the day before I was to start and show him my bruised and battered arms—I had 6-inch-long bruises where my IV had infiltrated and 3-inch-long bruises on the other arm from my preoperative bloodwork—to make sure he knew I wasn't a drug abuser. I couldn't explain to him what had happened and thankfully, he let it drop with "I had to have surgery." I didn't tell anyone at my new job what happened for over 6 months. There were still moments where it was very difficult. My first day on the job, I was filing student files and one of the students' middle names was Isabella, one of the names we had picked for a little girl. Many days I fought back tears and tried to hide my bruised hands from where I had beaten the floor or the counter in frustration.

Slowly, I began to come to terms with my fate. I did more research and found out that because I had had a partial molar pregnancy and my HCG dropped back to normal on its own within 7 weeks, I could start trying to conceive again in 3 months. I took charge of my own medical care and refused to have blood drawn every week. Because my veins were so horrible, I only had tests done at 4 weeks post D&C and then again 3 weeks later. At 7 weeks after surgery I was at less than 5 and that was good enough for me. Since I had decided I was only going to wait 3 months before trying to get pregnant again, and I don't like birth control pills anyway, we weren't using any protection. We weren't actively trying to get pregnant, but we weren't trying not to either. I had a period exactly 4 weeks after my surgery and then not for 6 weeks. I took a pregnancy test that month, just in case, but it was negative. Then I went to science camp with my son for a week. The end of that month is when I got pregnant. That was about 2½ months after my D&C. I honestly thought it would take longer for me to get pregnant and that it would be somewhere between 3 and 6 months. I was a little nervous calling the doctor to tell him I was pregnant again, because I knew he didn't want me to try for a year. It would take me almost to the end of the pregnancy to discover why he had insisted on a year wait.

My doctor wouldn't even see me until I was 8 weeks pregnant. He normally sees patients at 10 weeks and does the first ultrasound at 12 weeks, but because of my history, he did an ultrasound at that first visit. It was quite a relief to see the baby's heart beating! I was still nervous, though, because I had started spotting bright red at 6 weeks. Doing my own cervical checks had told me there was a soft spot of the top of my cervix and it seemed to be a bit dilated, so I was really worried about miscarriage. All my doctor would say was that he was 80%–90% sure everything would be fine, and if I was miscarrying there wasn't anything he could do anyway. As hard as it was, I just tried to stay positive and not worry too much, at least not out loud. Every little thing had me worrying, but I learned to keep it to myself. No one at work knew I was pregnant yet, nor did any of my family except my husband, daughter, and mom. We told my family after the 8-week ultrasound, and because my son went to the school where I worked he talked me into telling them when I was around 11 weeks.

The spotting and worrying continued intermittently throughout my first trimester and into the second, but the baby had a strong heartbeat, and everything else seemed to be going just fine. I tried to talk to my doctor about the soft spot on my cervix and my mounting suspicion that it was scar tissue from the D&C. He glossed over my questions every time, and I thought about switching doctors and hospitals many times. The fact that my previous labors had been so quick and the other hospital was almost half an hour away kept me where I was. When I was 29 weeks, I started spotting again after many weeks with no problems. I was also passing small blood clots and this worried me. My doctor left me wondering all day long before having his nurse practitioner call me back and tell me he was sure everything was fine. I had also told them I didn't think the baby was moving as much as normal. They didn't even offer to let me come in and listen to the baby's heartbeat. That was when I decided to switch doctors, no matter how far away the hospital was. I went to my last appointment with that doctor and had called ahead to get copies of my records. Looking through those records was hard to do but it also answered many of my questions.

I had requested my records for the molar pregnancy as well as the current pregnancy so my new doctor would have all the information. Looking at the surgical reports brought back all those feelings of anger and sadness. It did explain a lot though. The report showed that during my D&C, they had used a clamp to hold the top part of my cervix, the same part that was softer and that I suspected had suffered trauma. That explained the spotting as well as the blood clots; they were my body shedding the remaining scar tissue so that my cervix could function during delivery. The file also explained why my doctor had insisted on a year wait. Despite the pathology report that said I had had a possible partial molar pregnancy, my doctor continued to treat me as though I had had a complete molar pregnancy. I can only assume he decided it was a complete molar pregnancy when there was no baby at my first ultrasound. Even though it was hard to relive that horrible experience I was glad to have some of my questions answered. I was very happy with my new doctor and delivered a very healthy baby boy at almost 39 weeks. Baby Dominic (Nico) was 8 lb, 12 oz and perfect in every way. He is now a healthy, active 4-month-old, and I can't imagine life without him.

Although I would never want anyone to go through this terrible experience, because of it I had (and continue to have) an appreciation for my baby that I didn't have with my other kids. With my other babies there was a certain amount of taking it for granted. I enjoyed them but I didn't truly appreciate the miracles that they were. With Nico, sometimes I just look at him and can't believe he is really here. He absolutely takes my breath away. I enjoy every moment with him and wouldn't trade it for the world. Nothing takes away the worry, though. Being pregnant after a molar pregnancy is very nerve wracking. I worried my way through the entire pregnancy. The first few weeks I worried he would fall out, then I worried the next 12 weeks that my cervix would prove to be incompetent. The rest of the pregnancy I worried that my cervix wouldn't open when it needed to. During labor, when it became clear none of those fears had come true, I worried about his health. I watched every contraction on the screen and held my breath every time his heart rate dipped. Now that he is here, I only have the normal parent worries, but I know that if I ever were to be pregnant again I would worry all over again. One of the worst things about molar pregnancy is that it takes away that innocence that allows most women to enjoy their pregnancy. Women who have suffered a molar pregnancy know the worst can happen and spend the whole 9 months waiting to hear that it has.

Wendy

My husband and I had tried to conceive for a year. We had just started our path of infertility testing. I went in and had an HSG (to check out your fallopian tubes for blockages), and I was told that most of the time women conceive after the procedure because it essentially cleans out the cobwebs in the tubes. To my complete and utter delight, about 5 weeks later I was pregnant. I knew the day my period was due. I had been testing my ovulation and knew the exact day that I conceived. I was so excited; I had woken up at 6 A.M. and tested, and within seconds it was positive. After waking my husband I called my Mom. I told everyone imaginable. I had been waiting for this so long. I went in for bloodwork that Monday. My HCG level was 250, with progesterone of 40 (twice the normal amount). Two days later we checked my HCG level again, and it was in the 500s. I was so happy and was told that everything looked great. I wouldn't need my first obstetric appointment until about 10 weeks, but I couldn't wait. I pestered the nurses until I they gave me an appointment at about 8 weeks.

The appointment went well; we couldn't hear a heartbeat, but I was told it was too early. We waited a week and tried again, but it was still too early. In week 10 I went in and heard a heartbeat. I cried. But I was measuring big for the date, and we heard the heartbeat in two places on my belly, so we scheduled an ultrasound for the next day. I was thinking, Oh my God, I am having twins! I knew I was big for my dates. I couldn't wear any of my old clothes. I was already in baggy pants and was uncomfortable. I felt like I was carrying heavy water balloons in my uterus (later I found out that I really was—10 inches of cysts). The next day my husband was unable to get off work, so I went in alone, thinking we would find out we were having twins. WRONG. After 10 minutes, the technician went out to get the doctor. I knew something was very wrong and had already started crying. I told them they were scaring me and to tell me what was wrong. The doctor wasn't sure. It looked like the baby might have Down syndrome, and my placenta was very thick. What did this mean? The doctor was very reassuring and comforting, although I had never seen him before. He asked if I wanted to call my husband. They left the room and I cried hysterically. I couldn't reach my husband on the phone. I just remember thinking: What am I going to do? How did this happen?

The technician came back in, gave me a hug, and told me they were referring me to a specialty group down in Boston, Massachusetts (I lived in New Hampshire). The Boston group dealt with this on a daily basis and would be able to give me a better picture. I would be getting a phone call later that day to tell me when the appointment was. I sat in the car for 20 minutes crying. I have never cried like that before. I finally told my husband over the phone and while still crying; he barely understood what I was saying. He was and is my rock. He told me not to worry until we went to Boston, but that didn't happen for another week. It was the worst week of my entire life, but at the same time it was good because it gave us time to prepare for the worst and the best. We explored every option possible, so we thought.

We finally got to Boston, and they did the ultrasound. I could see the screen, and there were two hands, two legs, a head, and a strong heartbeat. I thought, Oh thank God, the baby's normal. The technician left the room, and I looked to my husband for reassurance. He saw it too. The doctor came in and said the worst I never imagined. The fetus was not viable! It was most likely a triploid anomaly. The fetus probably had three complete sets of chromosomes and would never survive. I also had cysts in the placenta and the ovaries that were consistent with a partial molar pregnancy. A WHAT? This was bad from the start and only got worse. The doctor was very compassionate and offered to do an amniocentesis to confirm the diagnosis, but we would have to wait 2 weeks for the report. I couldn't do it. I couldn't carry our baby for 2 more weeks, knowing it was already gone. She called my doctor back in New Hampshire, and he said he would meet with us as soon as we got back. She filled him in with all the details, but when we arrived, the doctor I had seen for the first ultrasound was not available. I saw yet another new doctor, who went over what the doctors had said in Boston: partial molar pregnancy, not viable, triploid, termination, cancer. I wanted this over. We asked to schedule the D&C as soon as possible, but the next day was Thanksgiving, and they only did emergent surgeries; this was not considered emergent. So it would have to be Friday, the day after Thanksgiving.

Needless to say, Thanksgiving wasn't very thankful. I spent most of the time trying to figure out what just happened and what was wrong with the baby. I have to say that I am a registered nurse, so I understand the medical jargon, and it scared the crap out of me. I was in a serious health crisis and had lost my baby, even though there was a strong heartbeat. I was on an emotional roller coaster ride. We went in Friday morning, the day after Thanksgiving. I was 15 weeks pregnant and measured at 20 weeks. My belly looked cute, but it was all wrong and I knew it. I was the only patient there that day, and everyone was very compassionate and loving toward me. The only thing I wanted to know when I woke up was the sex. The doctor said it was too tiny to tell. I think he was trying to spare me more pain. (I later found out it was a girl from the pathology report, and it did make it worse; I spend days thinking what she would have looked like.) I found out that my HCG level on the morning of surgery was over 1 million, and because of that I was now having hyperthyroidism, which can happen with HCG that high. I got to skip my first weekly blood draw because my HCG was too high to count in the hospital, but I then went into weekly HCG-level testing. It took forever to come down: a total of 24 weeks.

While dealing with this, I had another ultrasound to follow up on the ovarian cyst. I had a cyst that totaled about a 10-inch grapefruit in my belly. Not only was I having weekly blood draws, but I was having regular ultrasounds. I spent one day just crying. I didn't want any of this anymore. I wanted to be normal again. Little did I know I had been changed forever. I went out that weekend and bought regular jeans; I didn't want anything maternity. I still had a belly from the cysts and nothing fit; I finally found a pair of jeans two sizes bigger than my usual, but I didn't care. I wanted regular jeans. My husband was great. He tolerated everything and was wonderful throughout this whole ordeal. You not only have to deal with the loss but then you get to have a daily, or should I say weekly, reminder of the event that happened.

In the meantime, my husband got a new job, and we would be moving out of the country down to an island. I had to make sure that the doctors there were capable and aware of what I was going through. If they weren't, I was going to stay in the United States with my Mom until I was in the clear. I met another new doctor, and she was great. She had her training at Tufts University and had dealt with partial molar pregnancies before. It was February when we moved. My bloodwork continued, and I slowly began to feel better. I was allowed to go to monthly blood draws in April, and that was very hard after 5 months of weekly tests. I felt very insecure. I was scared to death after 2 weeks and called to have a level drawn. My doctor agreed, and my HCG was 11, which was considered negative in the islands. I had my doctor's appointment in May, and everything looked good.

In November 2001, I was cleared. We started trying to conceive in January 2003 with no luck. I went to my obstetrician in February, who suggested that I might not be ovulating regularly. He gave me a prescription for Clomid. I took the first round in March, and we conceived. I have to say it was one of the best days of my life, but also the scariest considering all that we went through. We were followed closely. It was a huge emotional roller coaster ride every day. We waited for the worst to happen again. The pregnancy went well, although at week 36 I developed pregnancy-induced hypertension. Finally, after 3 weeks of bed rest, my midwife induced me. We started the induction on Friday, December 5, 2003, and my little girl arrived December 6, 2003, at 5:27 P.M. Catherine Victoria is perfect in every way and was worth the events that led up to her birth. I can't imagine what my life would be like without her now.

I know all women wonder if it will ever happen, and how they would feel. Now I am another person to have a good ending after the molar pregnancy horror. I hope my story offers much hope to those who are on the path that I once was. Believe me, a year wait seems long, but it is well worth it once you get to hold your most precious gift ever. Having gone through what we all have gone through, you will appreciate your gift even more.

Best of luck to all and much baby dust when your time comes—and it will come. I only hope my story helps those who have been through a similar situation and who feel alone. I know I felt that no one would ever be able to understand, and for the most part that was true, except for those of us who have taken the time to share our stories and help others. Thank you to all of those who have shared. There is a light at the end of the tunnel and it shines brightly. It is our little angels who are in a better place together!

Appendix

Web Sites

MyMolarPregnancy.com
http://www.mymolarpregnancy.com
This is my Web site, founded in 2001. The site includes information and links related to molar pregnancy, a support group, and additional personal stories.

American Pregnancy Association: Molar Pregnancy
http://www.americanpregnancy.org/pregnancy
complications/molarpregnancy.html

Charing Cross Hospital Trophoblast Disease Service (London)
http://www.hmole-chorio.org.uk

Grief Loss and Recovery
http://grieflossrecovery.com/grief

Gynecologic Cancer Foundation
http://www.thegcf.org

HAND: Helping After Neonatal Death
http://www.handonline.org

Hannah's Prayer Ministries: Christian Support for Fertility Challenges
http://www.hannah.org

HopeXchange
http://www.hopexchange.com

International Society for the Study of Trophoblastic Diseases
http://www.isstd.org/index.html

Memorial Sloan-Kettering Cancer Center: Gestational Trophoblastic Disease
http://www.mskcc.org/mskcc/html/1909.cfm

MEND: Mommies Enduring Neonatal Death
http://home.mend.org

Miscarriage Support Auckland (New Zealand)
http://www.miscarriagesupport.org.nz/molar.html

Molarpregnancy.co.uk (United Kingdom)
http://molarpregnancy.co.uk/index.html

National Cancer Institute
http://www.cancer.gov

Personal Look at Molar Pregnancy
http://www.angelfire.com/nc2/mybabygirlnina

Secret Club Project: Understanding Pregnancy Loss Through the Arts
http://www.secretclubproject.org/home.html

Silent Grief
http://www.silentgrief.com

Women's Cancer Center
http://www.womenscancercenter.com/

Women's Cancer Network
http://www.wcn.org

Support Groups

MyMolarPregnancy
http://health.groups.yahoo.com/group/
mymolarpregnancy/

Choriocarcinoma
http://health.groups.yahoo.com/group/Choriocarcinoma

Eyes on the Prize: Support for Women With Gynecological Cancers
http://www.eyesontheprize.org

Molar Pregnancy Support Forum
http://molarpregnancy.co.uk/forum2 (United Kingdom)

MolarPregnancy2
http://health.groups.yahoo.com/group/molarpregnancy2

MP Moms
http://groups.yahoo.com/group/mpmoms
Group for women who have had successful pregnancies either before or after their molar diagnosis.

New Beginnings After Molar Pregnancy
http://health.groups.yahoo.com/group/Newbeginnings aftermolarpregnancy
Group for women trying to conceive again after a molar pregnancy.

Glossary

beta qual: Blood test that detects pregnancy based on the quality of the HCG in a blood sample. This is the blood test normally done when a pregnancy is suspected. It indicates the presence of elevated HCG but does not give a specific amount of HCG.

beta quant: Blood test used to detect the exact amount of HCG in women with gestational trophoblastic neoplasia. This test is used for nonroutine detection of HCG. "Normal" levels of HCG are generally considered to be levels less than 5, although this varies among physicians and facilities.

chemotherapy: The use of chemical agents to kill cancer cells or stop them from growing. These agents can be given intravenously or by mouth, depending on the drug being used. The agent most often used for a persistent mole is methotrexate.

choriocarcinoma: Cancerous form of gestational trophoblastic neoplasia that develops in placental tissue and may metastasize to other parts of the body. Usually curable if caught early enough and treated aggressively with chemotherapy.

dilation and curettage, D&C: A minor procedure in which the cervix is expanded enough to permit the cervical canal and the lining of the uterus to be scraped with an instrument known as a curette. This procedure is sometimes done after a miscarriage or during an abortion.

dilation and evacuation, D&E: Although essentially the same procedure as a D&C, a D&E is most often done for second-trimester miscarriages or abortions and uses more vacuum evacuation and requires more cervical dilation because of the larger quantity of tissue removed.

ectopic pregnancy: Pregnancy in which the fertilized egg implants not in the uterus but in the fallopian tube, ovary, or abdominal cavity. This is a serious condition and must be treated quickly.

gestational trophoblastic neoplasia: An "umbrella" term for any condition in women in which potentially cancerous cells develop in the placental tissues, including molar pregnancy and choriocarcinoma.

human chorionic gonadotropin, HCG: Hormone produced by the placenta that is detected by blood and urine pregnancy tests to indicate a pregnancy. HCG levels are affected by molar tissue, thus this hormone is used as an indicator for possible regrowth after a molar pregnancy.

hydatidiform mole: An abnormal growth of the membrane that encloses the embryo and gives rise to the placenta. If a mole develops, the embryo is usually either absent or dead. The mole itself is a collection of cysts that contain a jellylike substance and resemble a cluster of grapes. These cysts can grow very large if not removed, but most are removed by D&C. In a few cases, the mole can spread into the uterine muscle and cause bleeding. In very rare cases these moles can develop into choriocarcinoma.

intravenous, IV: As a verb, this term literally means "in the vein." The abbreviation IV is also used as a noun to refer to an intravenous line used to administer fluids.

invasive mole: Molar tissue that grows and spreads into and beyond the uterus, possibly metastasizing to other areas of the body.

molar pregnancy: An abnormality of the placenta caused by a problem when the egg and sperm join together at fertilization. A *complete molar pregnancy* occurs when sperm fertilizes an empty egg. No baby is formed, but placental tissue develops. The placenta grows rapidly and produces HCG. A *partial molar pregnancy* occurs when two sperm fertilize the same egg. Rather than developing into twins, something goes wrong, resulting in a pregnancy with an abnormal fetus and placenta. The fetus, which has too many chromosomes, usually dies in the uterus.

persistent mole: Molar tissue that regrows or continues to grow after the D&C. A second D&C may be performed before, or in addition to, treatment with chemotherapy.

ultrasound, abdominal: A painless, noninvasive procedure in which sound waves are used to produce images of the inside of the body. Reflected sound waves are received by instruments called transducers, which are small, hand-held devices that are moved back and forth across a patient's abdomen to form an image. A lubricating gel placed on the abdomen helps facilitate movement of the transducer.

ultrasound, vaginal: A form of ultrasound in which special probes are inserted into the vagina to obtain better images of a fetus or other uterine condition.

Bibliography

Armstrong D: Exploring fathers' experiences of pregnancy after a prior perinatal loss. *The American Journal of Maternal Child Nursing* 26(3):147–153, 2001

Batorfi J, Vegh G, Szepesi J, et al: How long should patients be followed after molar pregnancy? Analysis of serum HCG follow-up data. *European Journal of Obstetrics & Gynecology and Reproductive Biology* 112(1):95–97, 2004

Beaudoin CE, Tao CC: Benefiting from social capital in online support groups: an empirical study of cancer patients. *Cyberpsychology and Behavior* 10(4):587–590, 2007

Berkowitz RS, Goldstein DP: Presentation and management of molar pregnancy, in *Gestational Trophoblastic Disease*, 2nd Edition. Edited by Hancock BW, Newlands ES, Berkowitz RS. London, Chapman & Hall, 1997, pp. 127–142. Available online at http://www.isstd.org/gtd/index.html. Accessed October 13, 2007

Berkowitz RS, Im SS, Bernstein MR, Goldstein DP: Gestational trophoblastic disease: subsequent pregnancy outcome, including repeat molar pregnancy. *Journal of Reproductive Medicine* 43:81–86, 1998

Bruchim I, Kidron D, Amiel A, et al: Complete hydatidiform mole and a coexistent viable fetus: report of two cases and review of the literature. *Gynecologic Oncology* 77(1):197–202, 2000

Capitulo KL: Perinatal grief online. *The American Journal of Maternal Child Nursing* 29(5):305–311, 2004

Cote-Arsenault D, Bidlack D, Humm A: Women's emotions and concerns during pregnancy following perinatal loss. *The American Journal of Maternal Child Nursing* 26(3):128–134, 2001

Feltmate CM, Batorfi J, Fulop V, et al: Human chorionic gonadotropin follow-up in patients with molar pregnancy: a time for reevaluation. *Obstetrics & Gynecology* 101(4):732–736, 2003. Available online at http://www.greenjournal.org/cgi/reprint/101/4/732. Accessed October 11, 2007

Fishman DA, Padilla LA, Keh P, et al: Management of twin pregnancies consisting of a complete hydatidiform mole and normal fetus. *Obstetrics & Gynecology* 91:546–550, 1998

Garner EI, Lipson E, Bernstein MR, et al: Subsequent pregnancy experience in patients with molar pregnancy and gestational trophoblastic tumor. *Journal of Reproductive Medicine* 47:380–386, 2002

Garner E, Goldstein DP, Berkowitz RS, et al: Psychosocial and reproductive outcomes of gestational trophoblastic diseases. *Best Practice & Research. Clinical Obstetrics & Gynaecology* 17(6):959–968, 2003

Garner EI, Goldstein DP, Feltmate CM, et al: Gestational trophoblastic disease. *Clinical Obstetrics and Gynecology* 50(1):112–122, 2007

Growdon WB, Wolfberg AJ, Feltmate CM, et al: Postevacuation HCG levels and risk of gestational trophoblastic neoplasia among women with partial molar pregnancies. *Journal of Reproductive Medicine* 51(11):871–874, 2006

Icon Health Publications: *The Official Patient's Sourcebook on Gestational Trophoblastic Tumors: A Revised and Updated*

Directory for the Internet Age. Icon Health Publications, 2002

Icon Health Publications: *Molar Pregnancy: A Medical Dictionary, Bibliography, and Annotated Research Guide to Internet References.* Icon Health Publications, 2004

Im EO, Chee W, Lim HJ, et al: Patients' attitudes toward Internet cancer support groups. *Oncology Nursing Forum* 34(3):705–712, 2007

Johnson T, Schwartz M: *Gestational Trophoblastic Neoplasia: A Guide for Women Dealing With Tumors of the Placenta, such as Choriocarcinoma, Molar Pregnancy, and Other Forms of GTN.* Victoria, BC, Canada, Trafford Publishing, 2007

Kerkmeijer L, Wielsma S, Wiesma S, et al: Guidelines following hydatidiform mole: a reappraisal. *Australian and New Zealand Journal of Obstetrics and Gynaecology* 46(2):112–118, 2006

Kohn I: *A Silent Sorrow: Pregnancy Loss — Guidance and Support for You and Your Family, Revised and Updated 2nd Edition.* Oxford, England, Routledge, 2000

Kohorn EI: The new FIGO 2000 staging and risk factor scoring system for gestational trophoblastic disease: description and critical assessment. *International Journal of Gynecological Cancer* 11:73–77, 2001

Kurowski K, Yakoub N: Staying alert for gestational trophoblastic disease: implications for primary care clinicians. *Women's Health in Primary Care* 6:39–45, 2003. Available online at http://www.womenshealthpc.com/1_03/pdf/039GestTrophDisPC.pdf. Accessed October 18, 2007

Lanham CC: Pregnancy After a Loss: *A Guide to Pregnancy After a Miscarriage, Stillbirth, or Infant Death.* Berkley Trade, 1999

Lowdermilk D, Germino BB: Helping women and their families cope with the impact of gynecologic cancer. *Journal of Obstetric, Gynecologic, and Neonatal Nursing* 29:653–660, 2000

Matsui H, Sekiya S, Hando T, et al: Hydatidiform mole coexistent with a twin live fetus: a national collaborative study in Japan. *Human Reproduction (Oxford, England)* 15(3):608–611, 2000. Available online at http://humrep.oxfordjournals.org/cgi/reprint/15/3/608. Accessed October 11, 2007

Matsui H, Iitsuka Y, Suzuka K, et al: Subsequent pregnancy outcome in patients with spontaneous resolution of HCG after evacuation of hydatidiform mole: comparison between complete and partial mole. *Human Reproduction (Oxford, England)* 16:1274–1277, 2001. Available online at http:// humrep.oxfordjournals.org/cgi/reprint/ 16/6/1274. Accessed October 11, 2007

Matsui H, Iitsuka Y, Suzuka K, et al: Outcome of subsequent pregnancy after treatment for persistent gestational trophoblastic tumour. *Human Reproduction (Oxford, England)* 17(2):469–472, 2002. Available online at http://humrep.oxfordjournals.org/cgi/reprint/17/2/469. Accessed October 11, 2007

Olsen JH, Mellemkjaer L, Gridley G, et al: Molar pregnancy and risk for cancer in women and their male partners. *American Journal of Obstetrics and Gynecology* 181(3):630–634, 1999

Pezeshki M, Hancock BW, Silcocks P, et al: The role of repeat uterine evacuation in the management of persistent gestational trophoblastic disease. *Gynecologic Oncology* 95(3):423–429, 2004

Pisal N, Tidy J, Hancock B: Gestational trophoblastic disease: is intensive follow up essential in all women? *British Journal of Obstetrics and Gynaecology* 111(12):1449–1451, 2004

Royal College of Obstetricians and Gynaecologists: *The Management of Gestational Trophoblastic Neoplasia (Guideline No. 38).* London, U.K., Royal College of Obstetricians and Gyneacologists, 2004. Available online at http://www.rcog.org.uk/resources/Public/pdf/Gestational_Troph_Neoplasia_No38.pdf. Accessed October 11, 2007

Sebire NJ, Foskett M, Short D, et al: Shortened duration of human chorionic gonadotropin surveillance following complete or partial hydatidiform mole: evidence for revised protocol of a UK regional trophoblastic disease unit. *British Journal of Obstetrics and Gynaecology* 114(6):760–762, 2007

Seftel L: *Grief Unseen: Healing Pregnancy Loss Through the Arts.* Philadelphia, PA, Jessica Kingsley Publishers, 2006

Smith HO, Qualls CR, Prairie BA, et al: Trends in Gestational Choriocarcinoma: A 27-Year Perspective. *Obstetrics & Gynecology* 102:978–987, 2003. Available online at http://www.greenjournal.org/cgi/reprint/102/5/978. Accessed October 11, 2007

Steller MA, Genest DR, Bernstein MR, et al: Natural history of twin pregnancy with complete hydatidiform mole and coexisting fetus. *Obstetrics & Gynecology* 83:35–42, 1994

Stone H: *Forever Our Angels.* Morrisville, NC, Lulu.com, 2006

Tidy JA, Gillespie AM, Bright N, et al: Gestational trophoblastic disease: a study of mode of evacuation and subsequent need for treatment with chemotherapy. *Gynecologic Oncology* 78:309–312, 2000

Tuncer ZS, Bernstein MR, Goldstein DP, et al: Outcome of pregnancies occurring within 1 year of hydatidiform mole. *Obstetrics & Gynecology* 94(4):588–590, 1999. Available online at http://www.greenjournal.org/cgi/reprint/94/4/588. Accessed October 11, 2007

Wald HS, Dube CE, Anthony DC: Untangling the Web: the impact of Internet use on health care and the physician-patient relationship. *Patient Education and Counseling* 2007 [Epub ahead of print]

Wenzel LB, Berkowitz RS, Robinson S, et al: Psychological, social and sexual effects of gestational trophoblastic disease on patients and their partners. *Journal of Reproductive Medicine* 39(3):163–167, 1994

Wenzel LB: Psychosocial consequences of gestational trophoblastic disease, in *Gestational Trophoblastic Disease,* 2nd Edition. Edited by Hancock BW, Newlands ES, Berkowitz RS. London, U.K., Chapman & Hall, 1997, pp. 359–366. Available online at http://www.isstd.org/gtd/index.html. Accessed October 13, 2007

Wenzel L, Berkowitz RS, Newlands E, et al: Quality of life after gestational trophoblastic disease. *Journal of Reproductive Medicine* 47(5):387–394, 2002

Wolfberg AJ, Feltmate C, Goldstein DP, et al: Low risk of relapse after achieving undetectable HCG levels in women with complete molar pregnancy. *Obstetrics & Gynecology* 104(3):551–554, 2004. Available online at http://www.greenjournal.org/cgi/reprint/104/3/551. Accessed October 13, 2007

Wolfberg AJ, Berkowitz RS, Goldstein DP, et al: Postevacuation HCG levels and risk of gestational trophoblastic neoplasia in women with complete molar pregnancy. *Obstetrics & Gynecology* 106(3):548–552, 2005. Available online at http://www.greenjournal.org/cgi/reprint/106/3/548. Accessed October 11, 2007

Wolfberg AJ, Growdon WB, Feltmate CM, et al: Low risk of relapse after achieving undetectable HCG levels in women with partial molar pregnancy. *Obstetrics & Gynecology* 108(2):393–396, 2006. Available online at http://www.greenjournal.org/cgi/reprint/108/2/393. Accessed October 11, 2007